Natural Computing Series

T0181140

Martyn Amos

Theoretical and Experimental DNA Computation

With 78 Figures and 17 Tables

 Springer

Martyn Amos
Department of Computer Science
School of Engineering, Computer Science and Mathematics
University of Exeter
Harrison Building
Exeter EX4 4QF
United Kingdom
M.R.Amos@exeter.ac.uk
http://www.dcs.ex.ac.uk/~mramos

Series Editors
G. Rozenberg (Managing Editor)
rozenber@liacs.nl
Th. Bäck, J.N. Kok, H.P. Spaink
Leiden Center for Natural Computing
Leiden University
Niels Bohrweg 1
2333 CA Leiden, The Netherlands

A.E. Eiben
Vrije Universiteit Amsterdam
The Netherlands

ACM Computing Classification (1998): F.1–2, G.2.1–2, J.3

ISBN-13   978-3-642-08504-8   e-ISBN-13   978-3-540-28131-3

Springer is a part of Springer Science+Business Media

springeronline.com

© Springer-Verlag Berlin Heidelberg 2010
Printed in Germany

Cover Design: KünkelLopka, Werbeagentur, Heidelberg

Printed on acid-free paper   45/3142/YL – 5 4 3 2 1 0

This book is dedicated to the memory of Charlton Hall Amos.

This book is dedicated to the .... (Place at Will, flipped)

Preface

DNA computation has emerged in the last ten years as an exciting new research field at the intersection (and, some would say, frontiers) of computer science, biology, engineering, and mathematics. Although anticipated by Feynman as long ago as the 1950s [59], the notion of performing computations at a molecular level was only realized in 1994, with Adleman's seminal work [3] on computing with DNA. Since then the field has blossomed rapidly, with significant theoretical and experimental results being reported regularly.

Several books [120, 39] have described various aspects of DNA computation, but this is, to the author's best knowledge, the first to bring together descriptions of both theoretical and experimental results. The target audience is intentionally broad, including students as well as experienced researchers. We expect that users of the book will have some background in either computer science, mathematics, engineering, or the life sciences. The intention is that this book be used as a tutorial guide for newcomers to the field as well as a reference text for people already working in this fascinating area. To this end, we include two self-contained tutorial chapters (1 and 2), which convey only those aspects of computer science and biology that are required to understand the subsequent material.

We now describe in detail the structure of the book. An introduction places what follows in context, before we motivate the work to be presented by emphasizing some of the reasons for choosing DNA over other candidate computational substrates. One reason for choosing to work with DNA is the size and variety of the molecular "tool box" available to us. Since the discovery of the structure of DNA, a wide range of biological tools have been developed to facilitate and ease genetic experimentation. Molecular biologists have available to them a much more diverse set of methods for the manipulation of DNA than they have for any other molecule. It is possible that similar techniques could, in the future, become available for molecules other than DNA, but existing methods have the advantage of being ubiquitous, tried, and tested. We discuss these methods in more detail in Chap. 1.

In Chap. 2 we introduce some fundamental concepts in theoretical computer science and define what it means to "compute." We introduce the notion of a *computational problem*, and explain why some problems are fundamentally more difficult than others. This motivates Chap. 3, in which we show how models of molecular computation may be constructed. We describe several such formalizations that appeared soon after [3], and discuss their computational power.

One reason for the current interest in using DNA for computations is the massive parallelism inherent in laboratory operations on this particular molecule. When we perform a step in an experiment (say, adding an enzyme to a test-tube), the operation is performed *in parallel* on every molecule in the tube (in reality, reaction dynamics affect the overall efficiency of this process, but we discount this for the sake of clarity). If we consider that a single drop of solution can contain trillions of DNA molecules, the potential for *massive* parallelism is apparent. However, as we shall see, it is important to harness this parallelism correctly if we are to make significant progress. In Chap. 4 we discuss complexity issues in DNA computing, and outline the shortcomings of early models in the light of these.

By describing attempts to characterize the complexity of molecular algorithms in Chap. 4, we motivate a discussion of *feasible* and *efficient* models of DNA computation. We describe several models that attempt to drive the field closer to the so-called "killer application," the application of DNA-based computers that would establish their superiority in a particular domain.

Chapter 5 is concerned with physical realizations of some of the models outlined in Chap. 3. We describe several laboratory experiments and the lessons to be derived from the results obtained. We also describe exciting "late-breaking" laboratory results that appeared too recently to deal with in detail.

In Chap. 6 we discuss recent work that has focused attention on the potential for performing computations in vivo, as opposed to in vitro, as has previously always been the case. We examine the inner workings of the cell from a computational perspective, and describe recent theoretical and experimental developments.

Core work reported in this monograph was performed by the author as a doctoral student under the supervision of Alan Gibbons. I owe much to, and am grateful for, Alan's wisdom, generosity, patience, and humor. I thank Grzegorz Rozenberg for offering me the opportunity to publish this work in the Natural Computing series, and for his friendship and guidance. Many other colleagues have contributed to the work presented, especially Paul E. Dunne, David Hodgson, Gerald Owenson, and Steve Wilson. Chris Tofts was the external examiner of my doctoral thesis, and offered useful advice on content and presentation. I thank Cheryl Sutton for her help with final typesetting, and the assistance provided by Dr. Hans Wössner, Ingeborg Mayer, and Ronan Nugent at Springer, and by Christina Brückner at LE-Tex, is also gratefully acknowledged. Finally, I would like to express my gratitude for continued support from my wife, Justine Ashby.

Exeter   Martyn Amos
                                                      December, 2004

Contents

Introduction 1

1   DNA: The Molecule of Life 5
  1.1   Introduction 5
  1.2   The Structure and Manipulation of DNA 6
  1.3   DNA as the Carrier of Genetic Information 7
  1.4   Operations on DNA 10
  1.5   Summary.. 21
  1.6   Bibliographical Notes 21

2   Theoretical Computer Science: A Primer 23
  2.1   Introduction 23
  2.2   Algorithms and Automata 25
  2.3   The Turing Machine 27
  2.4   The Random Access Machine 29
  2.5   Data Structures 33
  2.6   Computational Complexity 39
  2.7   P and NP 43
  2.8   Summary.. 43
  2.9   Bibliographical Notes 44

3   Models of Molecular Computation 45
  3.1   Introduction 45
  3.2   Filtering Models....................................... 46
  3.3   Splicing Models 60
  3.4   Constructive Models 61
  3.5   Membrane Models 63
  3.6   Summary.. 69
  3.7   Bibliographical Notes 70

4   Complexity Issues 71
    4.1   Introduction 71
    4.2   An Existing Model of DNA Computation 73
    4.3   A Strong Model of DNA Computation 76
    4.4   Ogihara and Ray's Boolean Circuit Model 77
          4.4.1 Ogihara and Ray's Implementation 79
    4.5   An Alternative Boolean Circuit Simulation 82
    4.6   Proposed Physical Implementation 84
    4.7   Analysis 87
    4.8   Example Application: Transitive Closure 88
    4.9   P-RAM Simulation 90
    4.10 The Translation Process 94
    4.11 Assessment 100
    4.12 A Worked Example: The List Ranking Problem 102
    4.13 Summary 106
    4.14 Bibliographical Notes 107

5   Physical Implementations 109
    5.1   Introduction 109
    5.2   Implementation of Basic Logical Elements 109
    5.3   Initial Set Construction Within Filtering Models 110
    5.4   Adleman's Implementation 112
    5.5   Evaluation of Adleman's Implementation 115
    5.6   Implementation of the Parallel Filtering Model 117
    5.7   Advantages of Our Implementation 118
    5.8   Experimental Investigations 119
    5.9   Other Laboratory Implementations 135
          5.9.1 Chess Games 136
          5.9.2 Computing on Surfaces 138
          5.9.3 Gel-Based Computing 140
          5.9.4 Maximal Clique Computation 141
          5.9.5 Other Notable Results 143
    5.10 Summary 145
    5.11 Bibliographical Notes 145

6   Cellular Computing 147
    6.1   Introduction 147
    6.2   Successful Implementations 150
    6.3   Gene Unscrambling in Ciliates 150
    6.4   Biological Background 151
          6.4.1 IESs and MDSs 151
          6.4.2 Scrambled Genes 152
          6.4.3 Fundamental Questions 152

6.5   Models of Gene Construction 153
6.6   Summary 155
6.7   Bibliographical Notes 156

References 157

Index 167

Introduction

"Where a calculator on the ENIAC is equipped with 18,000 vacuum tubes and weighs 30 tons, computers in the future may have only 1,000 vacuum tubes and perhaps weigh $1\frac{1}{2}$ tons." – Popular Mechanics, 1949 [74]

This statement, made just over fifty years ago, is striking because it falls so short of reality. We have made huge advances in miniaturization since the days of room-sized computers, and yet the underlying computational framework (the *von Neumann architecture*) has remained constant. Today's supercomputers still employ the kind of sequential logic used by the mechanical dinosaurs of the 1930s.

There exist two main barriers to the continued development of "traditional", silicon-based computers using the von Neumann architecture. One is inherent to the machine architecture, and the other is imposed by the nature of the underlying *computational substrate*. A computational substrate may be defined as *"a physical substance acted upon by the implementation of a computational architecture."* Before the invention of silicon integrated circuits, the underlying substrates were bulky and unreliable. Of course, advances in miniaturization have led to incredible increases in processor speed and memory access time. However, there is a limit to how far this miniaturization can go. Eventually "chip" fabrication will hit a wall imposed by the *Heisenberg Uncertainty Principle* (HUP) (see [70] for an accessible introduction). When chips are so small that they are composed of components a few atoms across, quantum effects cause interference. The HUP states that the act of observing these components affects their behavior. As a consequence, it becomes impossible to know the exact state of a component without fundamentally changing its state.

The second limitation is known as the *von Neumann bottleneck*. This is imposed by the need for the central processing unit (CPU) to transfer instructions and data to and from the main memory. The route between the CPU and memory may be visualized as a two-way road connecting two towns. When the number of cars moving between towns is relatively small, traffic moves quickly. However, when the number of cars grows, the traffic slows down, and may even grind to a complete standstill. If we think of the cars as units of information passing between the CPU and memory, the analogy is complete. Most computation consists of the CPU fetching from memory and then executing one instruction after another (after also fetching any data required). Often, the execution of an instruction requires the storage of a result in memory. Thus, the speed at which data can be transferred between the CPU and memory is a limiting factor on the speed of the whole computer.

Some researchers are now looking beyond these boundaries and are investigating entirely new computational architectures and substrates. These include quantum, optical, and DNA-based computers. It is the last of these developments that this book concentrates on.

In the late 1950s, the physicist Richard Feynman first proposed the idea of using living cells and molecular complexes to construct "sub-microscopic computers." In his famous talk *"There's Plenty of Room at the Bottom"* [59], Feynman discussed the problem of "manipulating and controlling things on a small scale", thus founding the field of nanotechnology. Although he concentrated mainly on information storage and molecular manipulation, Feynman highlighted the potential for biological systems to act as small-scale information processors:

> The biological example of writing information on a small scale has inspired me to think of something that should be possible. Biology is not simply writing information; it is doing something about it. A biological system can be exceedingly small. Many of the cells are very tiny, but they are very active; they manufacture various substances; they walk around; they wiggle; and they do all kinds of marvelous things – all on a very small scale. Also, they store information. Consider the possibility that we too can make a thing very small which does what we want – that we can manufacture an object that maneuvers at that level! [59].

Since the presentation of Feynman's vision there there has been an explosion of interest in performing computations at a molecular level. Early developments, though, were theoretical in nature – the realization of performing computations at a molecular level had to wait for the development of the necessary methods and materials. In 1994, Adleman finally showed how a massively parallel random search may be implemented using standard operations on strands of DNA [3].

Previous proposals for molecular computers concentrated mainly on the use of proteins, but Adleman was inspired to use DNA by reading how the DNA polymerase enzyme "reads" and "writes" at the molecular level (for a personal account, see [2]). We describe Adleman's experiment in detail in Chap. 5, but for now note the similarity between the history of the development of "traditional" computers and that of their molecular counterparts.

Even though their underlying machine model (the von Neumann architecture) had been established for decades, the development of reliable electronic computers was only made possible by the invention of the transistor, which facilitated for the first time electronic manipulation of silicon. We may draw an interesting parallel between this series of events and the development of molecular computers. Although the concept dates back to the late 1950s, only now do we have at our disposal the tools and techniques of molecular biology required to construct prototype molecular computers. For example, just as the transistor amplifies *electrical* signals, the polymerase chain reaction (described in Chap. 1) amplifies *DNA* samples.

Adleman's experiment provided the field's foundations, but others quickly built upon them. It was clear that, while seminal, Adleman's approach was suitable only for the solution of a specific instance of a particular problem (the Hamiltonian Path Problem; see Chap. 5 for a full description of the experiment). If the idea of computing with molecules were to be applied to a range of *different* problems, then a general model of DNA computation was required, describing both the abstract operations available and their "real world" implementations. Several such models quickly appeared, each describing a framework for mapping high-level algorithmic descriptions down to the level of laboratory operations. These bridged the gap between theory and experiment, allowing practitioners to describe DNA-based solutions to computational problems in terms of both classical computer science and feasible biology. Theoretical models of DNA computation are described in depth in Chap. 3, with particular attention being given to the "mark and destroy" destructive model developed by the author and others.

The notion of feasibility, alluded to in the previous paragraph, means different things to different people. Whereas a biologist, working in an inherently "noisy" and unpredictable environment, may be prepared to accept a certain degree of error in his or her experimental results, a computer scientist may require absolute precision and predictability of results. Similarly, a computer scientist may only be interested in algorithms capable of solving problems of a size that cannot possibly be accomodated by even moderate leaps in the technology available to a bench biologist. This common disparity between expectation and reality motivates our study of the notion of the *complexity* of a molecular algorithm. We use the term not to describe the nature of the intricate interactions between components of an algorithm, but to derive some notion of the essential resources it requires. Computer scientists will be familiar with the concepts of time and (memory) space needed by a "traditional" algorithm, but it is also important to be able to describe these for molecular algorithms. Here, time may be measured in terms of the number of fundamental laboratory operations required (or the elapsed real time that they take to perform), and space in terms of the volume of solution needed to accommodate the DNA required for the algorithm to run. Studies of the complexity of DNA algorithms are described in detail in Chap. 4.

The initial rush of publications following Adleman's original paper was dominated mainly by theoretical results, but empirical results soon followed. In Chap. 5 we describe several laboratory implementations of algorithms within models mentioned in the previous paragraph. Most of the significant results are described in the abstract, in order to capture the essence of the experimental approach. In order to also highlight the factors to be considered when designing a protocol to implement a molecular algorithm, particular attention is given to the details of the experiments carried out by the author's collaborators.

With recent advances in biology, it is becoming clear that a genome (or complete genetic sequence of an organism) is not, as is commonly (and erroneously) suggested, a "blueprint" describing both components and their placement, but rather a "parts list" of proteins that interact in an incredibly complex fashion to *build* an organism. The focus of biology has turned toward reverse-engineering sequences of interactions in order to understand the fundamental processes that lead to life. It is clear that, in the abstract, a lot of these processes may be thought of in computational terms (for example, one biological component may act, for all intents and purposes, as a switch, or two components may combine to simulate the behavior of a logic gate). This realization has stimulated interest in the study of biological systems from a computational perspective. One possible approach to this is to build a computational model that captures (and, ultimately, predicts) the sequence of biological operations within an organism. Another approach is to view specific biological systems (such as bacteria) as reprogrammable biological computing devices. By taking well-understood genetic components of a system and reengineering them, it is possible to modify organisms such that their behavior corresponds to the implementation of some human-defined computation. Both approaches are described in Chap. 6.

In less than ten years, the field of DNA computation has made huge advances. Developments in biotechnology have facilitated (and, in some cases, been motivated by) the search for molecular algorithms. This has sometimes led to unfortunate speculation that DNA-based computers may, one day, supplant their silicon counterparts. It seems clear, however, that this unrealistic vision may one day be replaced by a scenario in which both traditional and biological computers coexist, each occupying various niches of applicability. Whatever the ultimate applications of biological computers may turn out to be, they are revolutionizing the interactions between biology, computer science, mathematics, and engineering. A new field has emerged to investigate the crossover between computation and biology, and this volume describes only its beginnings.

1

DNA: The Molecule of Life

"All rising to great places is by a winding stair." – *Francis Bacon*

1.1 Introduction

Ever since ancient Greek times, man has suspected that the features of one generation are passed on to the next. It was not until Mendel's work on garden peas was recognized (see [69, 148]) that scientists accepted that both parents contribute material that determines the characteristics of their offspring. In the early 20th century, it was discovered that *chromosomes* make up this material. Chemical analysis of chromosomes revealed that they are composed of both *protein* and *deoxyribonucleic acid*, or *DNA*. The question was, which substance carries the genetic information? For many years, scientists favored protein, because of its greater complexity relative to that of DNA. Nobody believed that a molecule as simple as DNA, composed of only four subunits (compared to 20 for protein), could carry complex genetic information.

It was not until the early 1950s that most biologists accepted the evidence showing that it is in fact DNA that carries the genetic code. However, the physical structure of the molecule and the hereditary mechanism was still far from clear.

In 1951, the biologist James Watson moved to Cambridge to work with a physicist, Francis Crick. Using data collected by Rosalind Franklin and Maurice Wilkins at King's College, London, they began to decipher the structure of DNA. They worked with models made out of wire and sheet metal in an attempt to construct something that fitted the available data. Once satisfied with their double helix model (Fig. 1.1), they published the paper [154] (also see [153]) that would eventually earn them (and Wilkins) the Nobel Prize for Physiology or Medicine in 1962.

1.2 The Structure and Manipulation of DNA

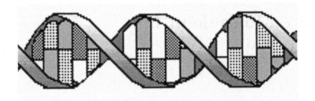

Fig. 1.1. Stylized depiction of DNA double helix

DNA (deoxyribonucleic acid) [1, 155] encodes the genetic information of cellular organisms. It consists of *polymer chains*, commonly referred to as DNA *strands*. Each strand may be viewed as a chain of *nucleotides*, or *bases*, attached to a sugar-phosphate "backbone." An n-letter sequence of consecutive bases is known as an n-mer or an *oligonucleotide*[1] of length n.

The four DNA nucleotides are adenine, guanine, cytosine, and thymine, commonly abbreviated to A, G, C, and T respectively. Each strand, according to chemical convention, has a 5' and a 3' end; thus, any single strand has a natural orientation. This orientation (and, therefore, the notation used) is due to the fact that one end of the single strand has a free (i.e., unattached to another nucleotide) 5' phosphate group, and the other end has a free 3' deoxyribose hydroxl group. The classical double helix of DNA (Fig. 1.2) is formed when two separate strands bond. Bonding occurs by the pairwise attraction of bases; A bonds with T and G bonds with C. The pairs (A,T) and (G,C) are therefore known as *complementary* base pairs. The two pairs of bases form *hydrogen bonds* between each other, two bonds between A and T, and three between G and C (Fig. 1.3).

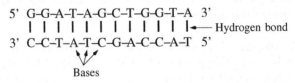

Fig. 1.2. Structure of double-stranded DNA

In what follows we adopt the following convention: if x denotes an oligo, then $\bar{x}$ denotes the complement of x. The bonding process, known as *annealing*,

---

[1] Commonly abbreviated to "oligo."

is fundamental to our implementation. A strand will only anneal to its complement if they have opposite polarities. Therefore, one strand of the double helix extends from 5' to 3', and the other from 3' to 5', as depicted in Fig. 1.2.

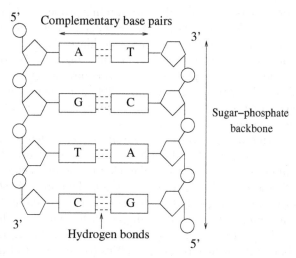

Fig. 1.3. Detailed structure of double-stranded DNA

1.3 DNA as the Carrier of Genetic Information

The *central dogma* of molecular biology [49] is that DNA produces RNA, which in turn produces proteins. The basic "building blocks" of genetic information are known as *genes*. Each gene codes for one specific *protein* and may be turned on (*expressed*) or off (*repressed*) when required.

Protein structure and function

Proteins are the working molecules in organisms, and the properties of living organisms derive from the properties of the proteins they contain. One of the most important functions proteins carry out is to act as *enzymes*. Enzymes act as specific *catalysts*; each type of enzyme catalyses a chemical reaction which would otherwise not take place at all, or at least take place very slowly. As an aside, the word "enzyme" is derived from the Greek for "in yeast", as all early work on enzymes was carried out on extracts of yeast [131]. The name of an enzyme indicates the type of reaction that it catalyses; for example, *restriction ligase* (see Sect. 1.4) catalyses the *ligation* of DNA strands (this process is described later). An organism's *metabolism* is defined as the the totality of the thousands of different chemical reactions occurring within it, catalysed by thousands of different enzymes. Proteins also have many other different functions, such as messengers and structural components (human hair is made up of the protein keratin). So, what determines their specific properties?

This question may be answered thus: a protein's properties result from the sequence of *amino acids* that comprise it. Proteins are linear chains of amino acids, strung together rather like beads on a necklace. There are 20 different amino acids, and, given that proteins can be anything from 50 to 500 amino acids in length, the number of possible proteins is beyond astronomical. If we assume that the average protein is made up of 300 amino acids, there are 20^{300} possible protein sequences. This dwarfs the estimated number of fundamental particles in the observable universe (10^{80} [131]).

We now concern ourselves with how protein sequence determines form, or structure. Each amino acid in the chain has a particular pattern of attraction, due to its unique molecular structure. The chain of amino acids *folds* into a specific three-dimensional shape, or *conformation*. The protein *self-assembles* into this conformation, using only the "information" encoded in its sequence. Most enzymes assemble into a globular shape, with cavities on their surface. The protein's *substrate(s)* (the molecule(s) acted upon during the reaction catalysed by that protein) fit into these cavities (or *active sites*) like a key in a lock. These cavities enable proteins to bring their substrates close together in order to facilitate chemical reactions between them.

Hemoglobin is a very important protein that transports oxygen in the blood. A visualisation[2] of the hemoglobin protein [121] is depicted in Fig. 1.4. A particularly important feature of the protein is the labelled active site, where oxygen binds.

When the substrate binds to the active site the conformation of the protein changes. This can be seen in Fig. 1.5, which depicts the formation of a hexokinase-glucose complex. The hexokinase is the larger molecule; on the left a molecule of glucose is approaching an active site (the cleft) in the protein. Note that the conformation of the enzyme changes after binding (depicted on the right). This behavior, as we shall see later, may be used in the context of performing computations.

There are many modern techniques available to determine the structure and sequence of a given protein, but the problem of predicting structure from sequence is one of the greatest challenges in contemporary bioinformatics. The rules governing the folding of amino acid chains are as yet not fully understood, and this understanding is crucial for the success of future predictions.

Transcription and translation

We now describe the processes that determine the amino acid sequence of a protein, and hence its function. Note that in what follows we assume the

---

[2] All visualisations were created by the author, using the RasMol [138] molecular visualisation package available from http://www.umass.edu/microbio/rasmol/

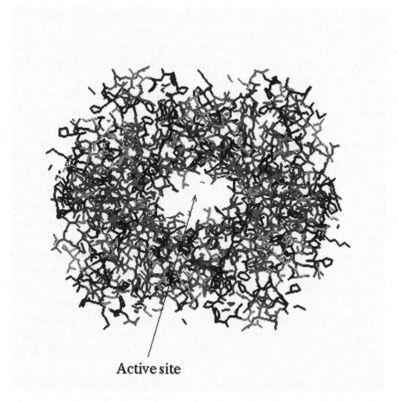

Active site

Fig. 1.4. Visualization of the hemoglobin protein

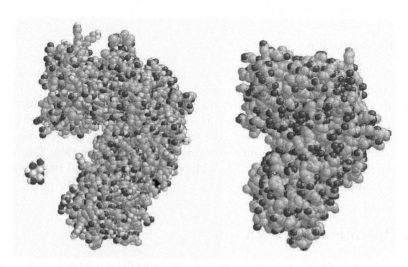

Fig. 1.5. Formation of a hexokinase-glucose complex

processes described occur in bacteria, rather than in higher organisms such as humans. In order for a DNA sequence to be converted into a protein molecule, it must be read (*transcribed*) and the transcript converted (*translated*) into a protein. Transcription of a gene produces a *messenger RNA* (mRNA) copy, which can then be translated into a protein.

Transcription proceeds as follows. The mRNA copy is synthesized by an enzyme known as *RNA polymerase*. In order to do this, the RNA polymerase must be able to recognize the specific region to be transcribed. This specificity requirement facilitates the regulation of genetic expression, thus preventing the production of unwanted proteins. Transcription begins at specific sites within the DNA sequence, known as *promoters*. These promoters may be thought of as "markers", or "signs", in that they are not transcribed into RNA. The regions that *are* transcribed into RNA (and eventually translated into protein) are referred to as *structural* genes. The RNA polymerase recognizes the promoter, and transcription begins. In order for the RNA polymerase to begin transcription, the double helix must be opened so that the sequence of bases may be read. This opening involves the breaking of the hydrogen bonds between bases. The RNA polymerase then moves along the DNA *template* strand in the $3 \rightarrow 5'$ direction. As it does so, the polymerase creates an *antiparallel* mRNA chain (that is, the mRNA strand is the equivalent of the Watson-Crick complement of the template). However, there is one significant difference, in that RNA contains uracil instead of thymine. Thus, in mRNA terms, "U binds with A."

The RNA polymerase moves along the DNA, the DNA re-coiling into its double-helix structure behind it, until it reaches the end of the region to be transcribed. The end of this region is marked by a *terminator* which, like the promoter, is not transcribed.

Genetic regulation

Each step of the conversion, from stored information (DNA), through mRNA (messenger), to protein synthesis (effector), is itself catalyzed by effector molecules. These effector molecules may be enzymes or other factors that are required for a process to continue (for example, sugars). Consequently, a loop is formed, where products of one gene are required to produce further gene products, and may even influence that gene's own expression. This process was first described by Jacob and Monod in 1961 [82], and described in further detail in Chap. 6.

1.4 Operations on DNA

Some (but not all) DNA computations apply a specific sequence of biological operations to a set of strands. These operations are all commonly used by molecular biologists, and we now describe them in more detail.

Synthesis

Oligonucleotides may be synthesized to order by a machine the size of a microwave oven. The synthesizer is supplied with the four nucleotide bases in solution, which are combined according to a sequence entered by the user. The instrument makes millions of copies of the required oligo and places them in solution in a small vial.

Denaturing, annealing, and ligation

Double-stranded DNA may be dissolved into single strands (or *denatured*) by heating the solution to a temperature determined by the composition of the strand [35]. Heating breaks the hydrogen bonds between complementary strands (Fig. 1.6). Since a $G - C$ pair is joined by three hydrogen bonds, the temperature required to break it is slightly higher than that for an $A - T$ pair, joined by only two hydrogen bonds. This factor must be taken into account when designing sequences to represent computational elements.

Annealing is the reverse of melting, whereby a solution of single strands is cooled, allowing complementary strands to bind together (Fig. 1.6).

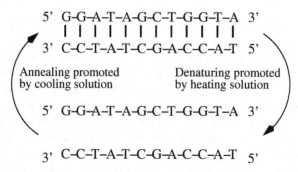

Fig. 1.6. DNA melting and annealing

In double-stranded DNA, if one of the single strands contains a discontinuity (i.e., one nucleotide is not bonded to its neighbor) then this may be repaired by DNA *ligase* [37]. This allows us to create a unified strand from several strands bound together by their respective complements. For example, Fig. 1.7a depicts three different single strands that many anneal, with a discontinuity where the two shorter strands meet. This may be repaired by the DNA ligase (Fig. 1.7b), forming a unified double-stranded complex (Fig. 1.7c).

Separation of strands

Separation is a fundamental operation, and involves the extraction from a test tube of any *single* strands containing a specific short sequence (e.g.,

(a)

5' G–G–A–T–A–G–C–T–G–G–T–A 3'

3' C–C–T–A–T–C 5'

3' G–A–C–C–A–T 5'

(b)

5' G–G–A–T–A–G–C–T–G–G–T–A 3'
   | | | | | | | | | | | |
3' C–C–T–A–T–C G–A–C–C–A–T 5'

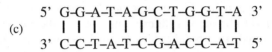

(c)

5' G–G–A–T–A–G–C–T–G–G–T–A 3'
   | | | | | | | | | | | |
3' C–C–T–A–T–C–G–A–C–C–A–T 5'

Fig. 1.7. (a) Three distinct strands. (b) Ligase repairs discontinuity. (c) The resulting complex

extract all strands containing the sequence $GCTA$). If we want to extract single strands containing the sequence x, we may first create many copies of its complement, $\overline{x}$. We attach to these oligos biotin molecules,[3] which in turn bind to a fixed matrix. If we pour the contents of the test tube over this matrix, strands containing x will anneal to the anchored complementary strands. Washing the matrix removes all strands that did not anneal, leaving only strands containing x. These may then be removed from the matrix.

Another removal technique involves the use of *magnetic bead separation.* Using this method, we again create the complementary oligos, but this time attach to them tiny magnetic beads. When the complementary oligos anneal to the target strands (Figure 1.8a), we may use a magnet to pull the beads out of the solution with the target strands attached to them (Fig. 1.8b).

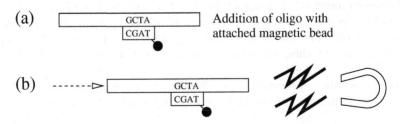

Fig. 1.8. Magnetic bead separation

---

[3] This process is referred to as "biotinylation".

Gel electrophoresis

Gel electrophoresis is an important technique for sorting DNA strands by size
[37]. Electrophoresis is the movement of charged molecules in an electric field.
Since DNA molecules carry a negative charge, when placed in an electric field
they tend to migrate toward the positive pole. The rate of migration of a
molecule in an *aqueous* solution depends on its shape and electric charge.
Since DNA molecules have the same charge per unit length, they all migrate
at the same speed in an aqueous solution. However, if electrophoresis is carried
out in a *gel* (usually made of agarose, polyacrylamide, or a combination of
the two), the migration rate of a molecule is also affected by its *size*.[4] This
is due to the fact that the gel is a dense network of pores through which the
molecules must travel. Smaller molecules therefore migrate faster through the
gel, thus sorting them according to size.

A simplified representation of gel electrophoresis is depicted in Fig. 1.9.
The DNA is placed in a well cut out of the gel, and a charge applied.

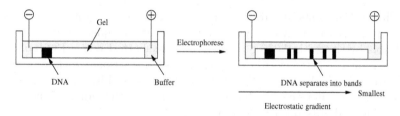

Fig. 1.9. Gel electrophoresis process

Once the gel has been run (usually overnight), it is necessary to visualize
the results. This is achieved by staining the DNA with the fluorescent dye
ethidium bromide and then viewing the gel under ultraviolet light. At this
stage the gel is usually photographed.

One such photograph is depicted in Fig. 1.10. Gels are interpreted as follows; each *lane* (1–7 in our example) corresponds to one particular sample
of DNA (we use the term *tube* in our abstract model). We can therefore run
several tubes on the same gel for the purposes of comparison. Lane 7 is known
as the *marker lane*; this contains various DNA fragments of known length, for
the purpose of calibration. DNA fragments of the same length cluster to form
visible horizontal *bands*, the longest fragments forming bands at the top of
the picture, and the shortest ones at the bottom. The brightness of a particular band depends on the amount of DNA of the corresponding length present
in the sample. Larger concentrations of DNA absorb more dye, and therefore

---

[4] Migration rate of a strand is inversely proportional to the logarithm of its molecular weight [114].

appear brighter. One advantage of this technique is its sensitivity – as little as 0.05 μg of DNA in one band can be detected as visible fluorescence.

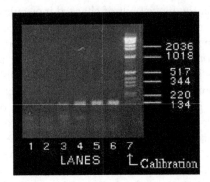

Fig. 1.10. Gel electrophoresis photograph

The size of fragments at various bands is shown to the right of the marker lane, and is measured in *base pairs* (b.p.). In our example, the largest band resolvable by the gel is 2,036 b.p. long, and the shortest one is 134 b.p. long. Moving right to left (tracks 6–1) is a series of PCR reactions which were set up with progressively diluted target DNA (134 b.p.) to establish the sensitivity of a reaction. The dilution of each tube is evident from the fading of the bands, which eventually disappear in lane 1.

Primer extension and PCR

The DNA *polymerases* perform several functions, including the repair and duplication of DNA. Given a short *primer* oligo, p in the presence of nucleotide triphosphates (i.e., "spare" nucleotides), the polymerase extends p if and only if p is bound to a longer *template* oligo, t. For example, in Fig. 1.11a, p is the oligo TCA which is bound to t, $ATAGAGTT$. In the presence of the polymerase, p is extended by a complementary strand of bases from the 5' end to the 3' end of t (Figure 1.11b).

Another useful method of manipulating DNA is the *Polymerase Chain Reaction*, or PCR [111, 112]. PCR is a process that quickly amplifies the amount of DNA in a given solution. Each cycle of the reaction doubles the quantity of each strand, giving an exponential growth in the number of strands.

PCR employs polymerase to make copies of a specific region (or *target sequence*) of DNA that lies between two *known* sequences. Note that this target sequence (which may be up to around 3,000 b.p. long) can be unknown ahead of time. In order to amplify template DNA with known regions (perhaps at either end of the strands), we first design forward and backward primers (i.e. primers that go from 5' to 3' on each strand. We then add a large excess (relative to the amount of DNA being replicated) of primer to the solution and heat

```
5' A T A G A G T T 3'
          | | |            (a)
3'        T C A   5'
```

```
5' A T A G A G T T 3'
   | | | | | | |           (b)
3' T A T C T C A   5'
```

Fig. 1.11. (a) Primer anneals to longer template. (b) Polymerase extends primer in the 5' to 3' direction

it to denature the double-stranded template (Fig. 1.12a). Cooling the solution then allows the primers to anneal to their target sequences (Fig. 1.12b). We then add the polymerase. In this case, we use Taq polymerase derived from the thermophilic bacterium *Thermus aquaticus*, which lives in hot springs. This means that they have polymerases that work best at high temperatures, and that are stable even near boiling point (Taq is reasonably stable at 94 degrees Celsius). The implication of this stability is that the polymerase need only be added once, at the beginning of the process, as it remains active throughout. This facilitates the easy automation of the PCR process, where the ingredients are placed in a piece of apparatus known as a *thermal cycler*, and no further human intervention is required.

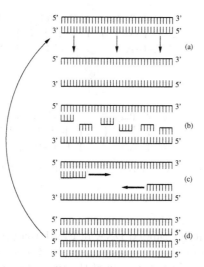

Fig. 1.12. (a) Denaturing. (b) Primer annealing. (c) Primer extension. (d) End result

The polymerase then extends the primers, forming an identical copy of the template DNA (Fig. 1.12c). If we start with a single template, then of course we now have two copies (Fig. 1.12d). If we then repeat the cycle of heating, annealing, and polymerising, it is clear that this approach yields an exponential number of copies of the template. A typical number of cycles would be perhaps 35, yielding (assuming a single template) around 68 billion copies of the *target sequence* (for example, a gene).

Unfortunately, the incredible sensitivity of PCR means that traces of unwanted DNA may also be amplified along with the template. We discuss this problem in a following chapter.

Restriction enzymes

Restriction endonucleases [160, page 33] (often referred to as *restriction enzymes*) recognize a specific sequence of DNA known as a *restriction site*. Any DNA that contains the restriction site within its sequence is cut by the enzyme at that point.[5]

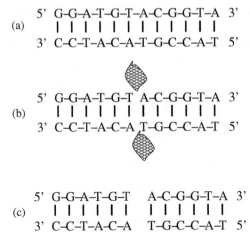

Fig. 1.13. (a) Double-stranded DNA. (b) DNA being cut by *Rsa*AI. (c) The resulting blunt ends

For example, the double-stranded DNA in Fig. 1.13a is cut by restriction enzyme *Rsa*I, which recognizes the restriction site *GTAC*. The enzyme breaks (or "cleaves") the DNA in the middle of the restriction site (Fig. 1.13b). The exact nature of the break produced by a restriction enzyme is of great importance. Some enzymes like *Rsa*I leave "blunt" ended DNA (Fig. 1.13c).

---

[5] In reality, only certain enzymes cut specifically at the restriction site, but we take this factor into account when selecting an enzyme.

Others may leave "sticky" ends. For example, the double-stranded DNA in Fig. 1.14a is cut by restriction enzyme *Sau*3AI, which recognizes the restriction site *GATC* (Fig. 1.14b). The resulting sticky ends are so-called because they are then free to anneal to their complement.

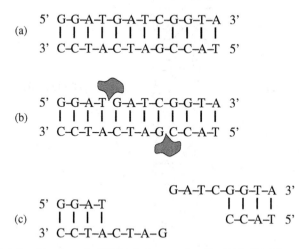

Fig. 1.14. (a) Double-stranded DNA being cut by *Sau*3AI. (b) The resulting sticky ends

Cloning

Once the structure of the DNA molecule was elucidated and the processes of transcription and translation were understood, molecular biologists were frustrated by the lack of suitable experimental techniques that would facilitate more detailed examination of the genetic material. However, in the early 1970s, several techniques were developed that allowed previously impossible experiments to be carried out (see [36, 114]). These techniques quickly led to the first ever successful cloning experiments [81, 102].

Cloning is generally defined as "... the production of multiple identical copies of a single gene, cell, virus, or organism." [130]. In the context of molecular computation, cloning therefore allows us to obtain multiple copies of specific strands of DNA. This is achieved as follows:

The specific sequence is inserted in a circular DNA molecule, known as a *vector*, producing a *recombinant DNA molecule*. This is performed by cleaving both the double-stranded vector DNA and the target strand with the *same* restriction enzyme(s). Since the vector is double stranded, restriction with suitable enzymes produces two short single-stranded regions at either end of the molecule (referred to as "sticky" ends. The same also applies to the target strand. The insertion process is depicted in Fig. 1.16. The vector and

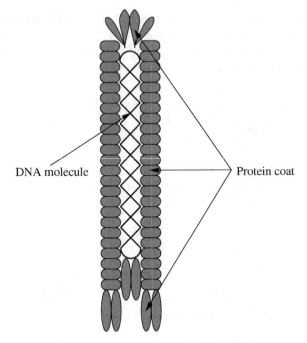

DNA molecule   Protein coat

Fig. 1.15. Schematic representation of the M13 phage structure

target are both subjected to restriction; then, a population of target strands is introduced to the solution containing the vector. The sticky ends of the target bind with the sticky ends of the vector, integrating the target into the vector. After ligation, new double-stranded molecules are present, each containing the new target sequence.

In what follows, we use the *M13 bacteriophage* as the cloning vector. Specifically, we use the M13mp18 vector, which is a 7,249 b.p. long derivative of M13 constructed by Yanisch-Perron et al. [164].

Bacteriophages (or *phages*, as they are commonly known) are viruses that infect bacteria. The structure of a phage is very simple, usually consisting of a single-stranded DNA molecule surrounded by a sheath of protein molecules (the *capsid*) (Fig. 1.15).

The vector acts as a *vehicle*, transporting the sequence into a *host* cell (usually a bacterium, such as *E.coli*). In order for this to occur, the bacteria must be made *competent*. Since the vectors are relatively heavy molecules, they cannot be introduced into a bacterial cell easily. However, subjecting *E.coli* to a variety of hot and cold "shocks" (in the presence of calcium, among other chemicals) allows the vector molecules to move through the cell membrane. The process of introducing exogenous DNA into cells is referred to as *transformation*. One problem with transformation is that it is a rather inefficient process; the best we can hope for is that around 5% of the bacterial cells will

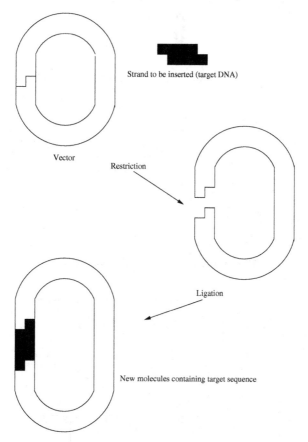

Fig. 1.16. Insertion of target strand into vector DNA

take up the vector. In order to improve this situation, we may use a technique known as *electroporation*. A high voltage pulse is passed through the solution containing the vectors and bacteria, causing the cell membranes to become permeable. This increases the probability of vector uptake. The vector then multiplies within the cell, producing numerous copies of itself (including the inserted sequence).

The infection cycle of M13 proceeds as follows. The phage attaches to a *pilus* (an appendage on the surface of the cell) and injects its DNA into the bacterium (Fig. 1.17a). The M13 DNA is not integrated into the DNA of the bacterium, but is still replicated within the cell. In addition, new phages are continually assembled within and released from the cell (Fig. 1.17b), which go on to infect other bacteria (Fig. 1.17c). When sufficient copies of the specific sequence have been made, the single-stranded M13 DNA may be retrieved from the medium. The process by which this is achieved is depicted in Fig. 1.18

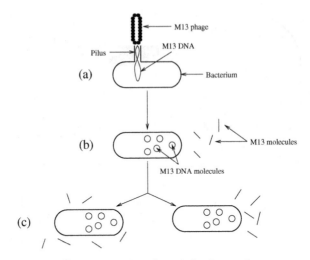

Fig. 1.17. M13 phage infection cycle

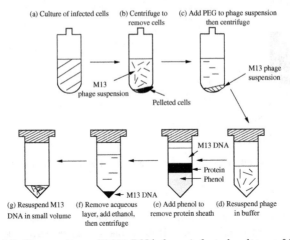

Fig. 1.18. Preparation of M13 DNA from infected culture of bacteria

(see also [104]). Once a sufficient volume of infected culture has been obtained we centrifuge it to pellet the bacteria (i.e., separate the bacteria from the phage particles). We then precipitate the phage particles with polyethylene glycol (PEG), add phenol to strip off the protein coats and then precipitate the resulting DNA using ethanol.

1.5 Summary

We described here the basic structure of DNA and the methods by which it may be manipulated in the laboratory. These techniques owe their origin to, and are being constantly improved by, the wide interests of molecular biologists working in modern areas such as the Human Genome project and genetic engineering. In Chap. 5 we show how these techniques allow us to implement a computation. Although other molecules (such as proteins) may be used as a computational substrate in the future, the benefit of using DNA is that this wide range of manipulation techniques is already available for use.

1.6 Bibliographical Notes

A definitive review of molecular genetics was co-authored by James Watson, one of the discoverers of the structure of DNA [156]. For in-depth information on molecular biology techniques, [18] is a laboratory manual that presents shortened versions of some 220 protocols selected from *Current Protocols in Molecular Biology*, the standard source in the field. For further details of the cloning process, the reader is directed to [136].

2

Theoretical Computer Science: A Primer

"Man is a slow, sloppy, and brilliant thinker; the machine is fast, accurate and stupid." – William M. Kelly

2.1 Introduction

A computer is a machine that manipulates information. The study of computer science is concerned with how this information is organized, manipulated, and used. However, what do we mean by *information*?

The basic unit of information is the *bit*, which can take one of two mutually exclusive values (for example, "true" or "false", or "on" or "off"). The fundamental components of digital computers use signals that are either "on" or "off." The details of computer architecture are beyond the scope of this book, but the interested reader is referred to [149].

Numbers are represented within computers using the *binary number system*. Within this system, the two symbols 0 and 1 are sufficient to represent any number (within the storage limits of the computer, obviously). Any integer value that is not a power of 2 can be expressed as the sum of two or more powers of 2. Each binary digit in a number is twice as significant as the digit to its right, and half as significant as the digit on its left. So, for example, the binary number 1011 corresponds to $(1 \times 2^3) + (0 \times 2^2) + (1 \times 2^1) + (1 \times 2^0) = 11$. In order to convert a decimal number to binary, we continually divide the number by two until the quotient is zero. The remainders, in reverse order, give the binary representation of the number.

The fundamental computational components introduced earlier are generally known as *gates*. These gates may be thought of as "black boxes" that take in signals and yield an output. This is known as *computing a function*. The three gates that we will initially consider are called NOT, AND, and OR. The conventional symbols used to depict these gates are depicted in Fig. 2.1, along with their functional behavior.

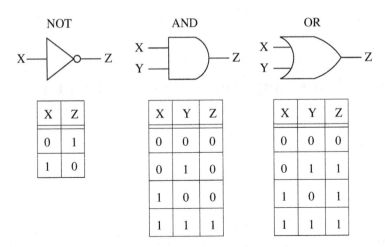

Fig. 2.1. NOT, AND and OR gates

Let us first look at the simplest of the three, the NOT gate. This gate takes a single input, X. If $X = 0$, then the output, Z, of the NOT gate is 1. Conversely, if $X = 1$ then $Z = 0$. Clearly, this gate "flips" its input, from 1 to 0 and vice versa. X and Z are both examples of *variables*, as their values may change.

The AND gate is more complex. Note that it outputs 1 if and only if *both* its inputs are equal to 1, and outputs 0 for all other input cases. This may be used, for example, in a situation where we wanted to say something like "if the washing machine has been turned off *and* the drum has stopped spinning then open the door." We could model this system with a single AND gate, where X is 1 only if the machine has been turned off, Y is 1 only if the drum has stopped spinning, and Z having the value 1 means that the door may be opened.

Now let us consider the OR gate. This outputs 1 if *any* of its inputs are equal to 1, and 0 otherwise. For example, we may want to express the following: "if it's raining *or* the trains are late, then drive to work." The OR gate could model this, where X is 1 only if it's raining, Y is 1 only if the trains are late, and Z is 1 if the decision is to drive to work. In this situation, if *either* X or Y is 1 then $Z = 1$.

Of course, these gates do not usually work in isolation. We may combine these gates into *circuits* to model more complex situations. In order to describe these circuits we need a new type of algebra, in which variables and functions can take only the values 0 and 1. This algebra is called a *Boolean algebra*, after its discoverer, George Boole.

Boolean algebra

A Boolean function has one or more input variables and yields a result that depends only on the values of those variables. One function may be defined by saying that $f(X)$ is 1 if X is 0, and $F(X)$ is 0 if X is 1. This is obviously the NOT function described above.

Because each Boolean function of n variables has only 2^n possible combinations of input values, we can describe the function completely by listing a table of 2^n rows, each specifying a unique set of input values with the corresponding value for the function. Such tables are known as *truth tables*, and the tables of Fig. 2.1 are examples of these.

We may combine gates to form *circuits* to evaluate more complicated functions. For example, we may want to say "if the car ignition is on and the seatbelt is not fastened then sound the buzzer." The circuit to implement this function is depicted in Fig. 2.2.

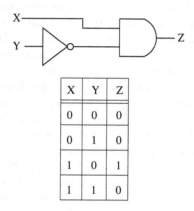

X	Y	Z
0	0	0
0	1	0
1	0	1
1	1	0

Fig. 2.2. Example circuit

Here, X represents the state of the ignition, and Y the state of the seatbelt. The buzzer sounds ($Z = 1$) only if X is 1 (i.e., the ignition is on) and Y is 0 (i.e., *not* 1, or the seatbelt is *not* on).

Boolean logic circuits implement computations by taking in inputs, applying some function, and producing an output. We now examine the nature of computation in more detail.

2.2 Algorithms and Automata

An *algorithm* is a mathematical procedure for performing a computation. To phrase it another way, an algorithm is "a computable set of steps to achieve a desired result." As an aside, we note that the term is derived from the name of the Persian author Abu Ja'far Mohammed ibn Mûsâ al-Khowârizmî who wrote a book detailing arithmetic rules in around 825 A.D. An algorithm is abstract; it is *not* a program (rather, a program is an *implementation* of an algorithm).

Algorithms are executed in a "mechanical" fashion (just as a computer mechanically executes a program). The mechanistic nature of algorithms allows us to introduce the abstract concept of a computing *machine*. Computing machines essentially take in some *input* and compute some *output*. Because of their abstract nature, we ignore "real-world" limitations imposed by, for example, memory size or communication delays.

A typical problem that we may pose a computing machine is to sort a list of integers into ascending order. Thus, the input to the machine is a list of integers (separated by spaces), and the output is also a list of integers. The problem is to compute a *function* that maps each input list onto its numerically ordered equivalent. Other problems involve data structures more complicated than simple lists, and we shall encounter these later.

The simplest computing machine is the *finite-state automaton*. We consider the *deterministic finite-state automaton* (DFA). The DFA simply reads in strings and outputs "accept" or "reject." This machine has a set of *states*, a *start state*, an input *alphabet*, and a set of *transitions*. One or more of the states are nominated as *accepting states*. The machine begins in the start state and reads the input string one character at a time, changing to new states determined by the transitions. When all characters have been read, the machine will either be in an accepting state or a rejecting state.

An example DFA is depicted in Fig. 2.3. This DFA has an alphabet of (a,b) and two states, 1 (the start state) and 2 (the only accepting state, denoted by the double circle). The transitions are denoted by the labels attached to the arrowed lines. For example, if the machine is in state 2 and it reads the character "b", the only available transition forces it into state 1.

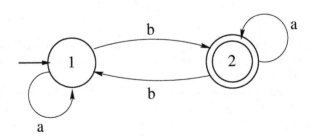

Fig. 2.3. Example depiction of a DFA

So, this DFA accepts strings containing only the characters "a" and "b", and accepts a string if and only if it contains an *odd* number of "b"s. Consider the following examples;

Example 1: Input string "abba". Start in state 1 and read "a". Stay in state 1. Read "b", move to state 2. Read "b", move to state 1. Read "a", stay in state 1. As the input is now exhausted, we check the current state and find that the machine rejects the input, as it has an *even* number of "b"s.

Example 2: Input string "babb". Start in state 1 and read "b". Move to state 2. Read "a", stay in state 2. Read "b", move to state 1. Read "b", move to state 2. As the input is now exhausted, we check the current state and find that the machine accepts the input, as it has an *odd* number of "b"s.

Rather than depicting state transition rules with a diagram, we can present them in the form of a table (Table 2.1).

Table 2.1. Rules for DFA

State	Symbol	New state
1	a	1
1	b	2
2	a	2
2	b	1

Although fundamental to the theory of computer science, finite-state automata are limited because, as their name suggests, they are of fixed size. This severely limits their functionality. We could not, for example, solve the sorting problem above for lists of *arbitrary* size using a fixed finite-state automaton. We require an "infinite" machine that is capable of solving problems of arbitrary size. The first such machine was the *Turing Machine*.

2.3 The Turing Machine

The Turing Machine (TM) was introduced by Alan Turing in 1936 [150]. This is, essentially, a finite automaton as described above, augmented with an unbounded external storage capacity. More specifically, the TM consists of a finite automaton controller, a read-write head, and an unbounded sequential tape memory. This tape memory is linear and consists of *cells*, each of which can contain one symbol (or be blank) (Fig. 2.4).

Depending on the current state of the machine and the symbol currently being read from the tape, the machine can change its state, write a symbol (or blank) to the tape, and move the head left or right. When the machine has no rule for a combination of state and symbol it *halts*.

The following example illustrates the operation of the TM. This machine has five states, the initial state being 1. The program (or set of rules) is

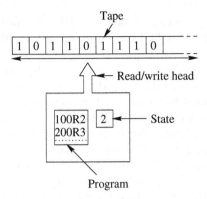

Fig. 2.4. Turing Machine

denoted in Table 2.2. The finite automaton controller may be thought of as a *program*, consisting of transition rules of the form "in state n, if the head is reading symbol x, write symbol y to the tape, move left or right one cell on the tape, and change to state m."

The table could be expressed by writing each row as a statement, substituting "L" for "Left", "R" for "Right", and "B" for "Blank." This gives the following TM program:

(1, 0, 0, R, 2)
(2, 0, 0, R, 3)
(2, 1, 1, R, 2)
(3, 0, B, L, 5)
(4, 0, 1, R, 2)

Table 2.2. Example Turing Machine program

State	Symbol	Write	Move	New state
1	0	0	Right	2
2	0	0	Right	3
2	1	1	Right	2
3	0	Blank	Left	5
3	1	0	Left	4
4	0	1	Right	2

The TM starts with its read-write head on the left-most cell. The input string is supplied on the tape, the first character of it occupying the left-most cell. The TM then repeatedly applies the supplied transition rules (or "runs" the program).

The example program given above implements *unary addition*. The input represents two numbers to be added, expressed in unary notation. To represent two integers $\{j, k\}$ as an input string we start with a marker "0", followed by j "1"s, followed by a separator "0", and then k "1"s terminated by a final "0." So, to add 2 and 3 we would specify an input string on 01101110.

Running the program specified in Table 2.2 on the input string 01101110 involves the steps described in Table 2.3. Note that the position of the read-write head on the tape is denoted by an arrow.

Table 2.3. Run of TM program

Tape	State	Action
0̌1101110	1	Write 0, go to state 2
01̌101110	2	Scan over 1
011̌01110	2	Scan over 1
0110̌1110	2	End of first number, go to next
01101̌110	3	Change 1 to 0, go back
0111̌0110	4	Copy 1, return to second number
011110̌10	2	End of first number, go to next
0111101̌0	3	Change 1 to 0, go back
0111101̌0	4	Copy 1, return to second number
0111110̌0	3	No second number, erase trailing 0
0111110̌	5	Halt

It is clear that, after the TM halts, the tape represents the correct sum represented in unary notation.

2.4 The Random Access Machine

Any function (problem) that can be computed by any other machine can also be computed by a Turing Machine. This thesis was established by Alonzo Church (see [145]). This means that Turing Machines are *universal* in the sense of computation. Note that this does not necessarily mean that *any* function can be computed by a Turing Machine – there exist *uncomputable* functions. The most famous uncomputable problem is the *Halting Problem*: given a Turing Machine with a given input, will it halt in a finite number number of steps or not? This can be re-phrased in the following fashion: is it possible to write a program to determine whether any *arbitrary* program will halt? The answer is no.

So, we can see that the Turing Machine is useful for investigating fundamental issues of computability. However, the ease of its application in other domains is less clear. Writing a program for a Turing Machine is a very "low-level" activity that yields an awkward and non-intuitive result. TMs also differ from "traditional" computers in several important ways, the most important concerning memory access. Recall the run-through of the example TM program, given in Table 2.3. In order to reach the end of the first number on the tape, the TM was forced to scan linearly along the tape until it reached the middle marker. In more general terms, in order to reach a distant tape (memory) cell, the TM must first read all of the intermediate cells.

This problem may be overcome by the introduction of the *Random Access Machine (RAM)* [5]. This machine is capable of reading an arbitrary memory cell in a single step. Like the TM, the RAM is also abstract in the sense that it has unbounded memory and can store arbitrarily large integers in each cell.

The structure of the RAM is depicted in Fig. 2.5. The main components of the machine are the read-only input tape, a write-only output tape, a fixed program, and a memory. The input and output tapes are identical in structure to the TM tape. When a symbol is read from the input tape, the read head is moved one cell to the right. When a symbol is written to the output tape, the write head is moved one cell to the right.

The RAM program is fixed, and is stored separately from the memory. The first memory location, M[0] is reserved for the *accumulator*, where all computation takes place.

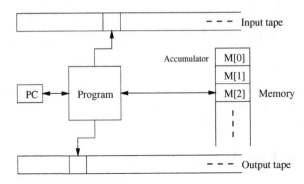

Fig. 2.5. Random Access Machine

In order to describe algorithms, we define an *instruction set* and a program. The instruction set defines the range of operations available to the RAM. The exact nature of the operations chosen is not important, and different authors specify different instruction sets. However, the set of instructions usually resemble those found in real computers. The program is simply a sequence of instructions, any of which may be *labelled*. The *program counter* (PC) keeps track of the current execution point in the program.

An example instruction set is given in Table 2.4. Instructions are of the form < *operation* >< *operand* >. An operand can be:

1. $= i$, indicating the integer i,
2. A nonnegative integer i, indicating the contents of memory location M[i], or
3. $*i$, indicating the contents of M[j], where j is the integer found in M[i]; This is known as *indirect addressing*.

Note that the $x \leftarrow y$ notation is interpreted as "x takes the value of y." For example, ACC $\leftarrow$ M[a] means "copy the contents of memory location M[a] into the accumulator." After [5], we define *value*(a), the value of an operand a, as follows:

1. *value*($= i$)$=i$,
2. *value*(i) $=$ M[i],
3. *value*($*i$) $=$ M[M[i]].

Table 2.4. Example RAM instruction set

Instruction	Meaning
LOAD a	ACC $\leftarrow$ *value*(a)
STORE i	M[i] $\leftarrow$ ACC
STORE $*i$	M[M[i]] $\leftarrow$ ACC
ADD a	ACC $\leftarrow$ ACC + *value*(a)
SUB a	ACC $\leftarrow$ ACC-*value*(a)
MULT a	ACC $\leftarrow$ ACC$\times$*value*(a)
DIV a	ACC $\leftarrow$ ACC$\div$*value*(a)
READ i	M[i] $\leftarrow$ current input symbol
READ $*i$	M[M[i]] $\leftarrow$ current input symbol
WRITE a	Print *value*(a) onto output tape
JUMP l	Set PC to the instruction labeled l
JGTZ l	If ACC >0, PC $\leftarrow$ instruction labeled l, else PC $\leftarrow$ next instruction
JZERO l	If ACC $= 0$, PC $\leftarrow$ instruction labeled l, else PC $\leftarrow$ next instruction
HALT	End execution

It should be clear that programs written using this instruction set are rather easier to understand than those written for a Turing Machine. Nontheless, for the purposes of algorithmic description we may abstract even further away from the machine level by introducing pseudo-code. Pseudo-code expresses the mechanisms underlying a program without necessarily tying that expression to any one particular machine or programming language. Pseudo-code shows program flow by using constructs such as "if some condition is true, then do procedure 1, else do procedure 2", "while some condition is true, do something", and "do something n times." These constructs are very common, and are found in the majority of programming languages.

We now give an example for the purposes of illustration. Consider the function $f(n)$, given by

$$f(n) = \begin{cases} 2^n & \text{if } n \geq 1 \\ 0 & \text{otherwise} \end{cases}$$

This function simply returns 2^n if n is greater than or equal to 1, and 0 otherwise. The following pseudo-code program computes $f(n)$:

```
read n
if n ≤ 0 then write 0
      else
      begin
          power ← 1
          while n > 0 do
              begin
                  power ← power × 2
                  n ← n − 1
              end
          write power
      end
end
```

The labels n, *power*, and *counter* are assigned to denote M[1], M[2], and M[3] respectively (these are referred to as *variables*, as their value may change during the execution of the program). Note that the program flow is depicted using **begin** and **end** to denote program *blocks*, and that blocks at different levels of control are indented differently.

We now give the corresponding RAM program in order to illustrate *the mapping* with the pseudo-code.

	READ 1		**read** n
	LOAD 1	↑	
	JGTZ *numpos*	‖	**if** $n \leq 0$ **then write** 0
	WRITE = 0	↓	
	JUMP *end*		
numpos:	LOAD 1	↑	
	STORE 2	↓	*power* ← n
	LOAD 1	↑	
	SUB = 1	‖	*counter* ← *n*-1
	STORE 3	↓	
while:	LOAD 3	↑	
	JGTZ *continue*	‖	**while** *counter* > 0 **do**
	JUMP *endwhile*	↓	
continue:	LOAD 2	↑	
	MULT = 2	‖	*power* ← *power* × 2
	STORE 2	↓	
	LOAD 3	↑	
	SUB = 1	‖	*counter* ← *counter* − 1
	STORE 3	↓	

	JUMP *while*	
endwhile:	WRITE 2	**write** *power*
end:	HALT	

The description of the structure of the RAM and its instruction set consti-
tutes the *machine model* or *model of computation*. This is a formal, abstract
definition of a computer. Using a model of computation allows us to calculate the resources (running time and memory space) required by algorithms,
without having to consider implementation issues. As we have seen, there are
many models of computation, each differing in computing power. In Chap.5
we consider models of computation using DNA.

2.5 Data Structures

So far, we have only considered very simple problems, such as computing
powers of integers. However, many "interesting" problems are concerned with
other mathematical constructs. The solution of a computational problem often involves the determination of some property (or properties) of a given
mathematical structure. Such structures include lists, networks, and trees.

These are examples of *data structures*. A data structure is a way of orga-
nizing information (usually, but not always, in computer memory) in a self-
contained and organized fashion. Data structures have algorithms associated
with them to access and maintain the information they contain. For example,
one of the simplest data structures is the *array*. Arrays are used when we need
to store and manipulate a collection of items of similar *type* that may have
different *attributes*. Consider a set of mailboxes, or "pigeon-holes." Each is
identical (i.e., of the same *type*), but may have different contents (*attributes*).
In addition, each mailbox has a unique number, or *address*. These principles
hold just as well if, for example, we wish to store a list of integers for use
within an algorithm. Rather than declaring an individual variable for each
integer stored, we simply declare an array of integers that is large enough for
our purposes.

Consider an example: we wish to store the average annual rainfall in our
country (rounded up to the nearest centimeter) over a ten-year period. We
could declare a separate variable for each year to store the rainfall over that
period, but this would be unwieldy and inefficient. A much better solution is
to declare an array of ten integers, each storing one year's rainfall. Thus, we
may declare an array called *rainfall*, addressing the elements as *rainfall*[0],
rainfall[1], ..., *rainfall*[9]. Note how the first element of an n-element array
always has address 0, and the last has the address n-1.

Arrays may be multi-dimensional, allowing us to represent more complex
information structures. For example, we may wish to write an algorithm to calculate the driving distances between several towns. The first problem we face
is how to represent the information about the towns and the roads connecting

them. One solution is the *adjacency matrix*, which is a two-dimensional array of integers representing the distances between towns. An example is shown in Fig. 2.6. Here, we have five towns connected by roads of a given length. We declare a two-dimensional array *distance*[5][5] to store the distance between any two towns.

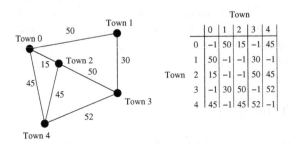

Fig. 2.6. Example map and its corresponding adjacency matrix

The distance is then found by reading the contents of array element *distance*[*firsttown*][*secondtown*]. Notice that for each town n, *distance*[n][n] has the value -1, as it makes no sense for a town to have a distance to itself. Also, if there is no direct road connecting town x and town y, then *distance*[x][y] also takes the value -1. The final thing to note about the adjacency matrix is that as the roads are two-way, the matrix is *symmetrical* in the diagonal (i.e., *distance*[x][y] has the same value as *distance*[y][x]). Obviously, this is not the most efficient method of storing *undirected* graphs, but it provides an easy illustration of the use of more complicated data structures.

Graphs

As we have seen, an example problem may be phrased thus: given a set of towns connected by roads, what is the shortest path between town A and town B? Another problem may ask if, given a map of mainland Europe, it is possible to color each country red, green, or blue such that no adjacent countries are colored the same.

The first example is referred to as an *optimisation* problem, as we are required to find a path that fits some criterion (i.e., it is the shortest possible path). The second example is known as a *decision* problem, as we are answering a simple "'yes/no" question about a structure. A large body of problems is concerned with the mathematical structures known as *graphs*. In this context, a graph is not a method of visualising data, rather a network of points and lines. More formally, a graph, $G = (V, E)$, is a set of points (*vertices*), V, connected by a set of lines (*edges*), E. Take the fragment of an English county map depicted in Fig. 2.7.

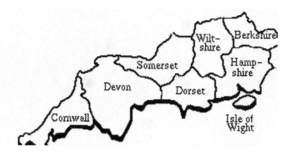

Fig. 2.7. Fragment of county map

The graph representing this map is depicted in Fig. 2.8, with vertices representing counties and edges representing borders.

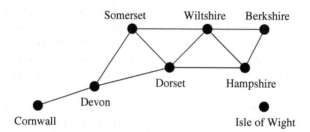

Fig. 2.8. Graph representing county map

Vertices and edges may have labels attached to them (for example, a vertex may be labelled with the name of a county, and an edge may be labelled with the length of the border.) Vertices may be isolated (i.e., be unconnected to any other vertex). For example, in Fig. 2.7, the Isle of Wight is an island, so the vertex representing it is isolated. In addition, edges may be *directed* or *undirected*. As their name suggests, the former have an implied direction (if, for example, they represent one-way streets or flights between two airports), while the latter have no particular direction attached to them.

Another graph is is presented in Fig. 2.9. As we denote a graph by $G = (V, E)$, where V is the vertex set and E the edge set, Fig. 2.9 represents the graph $G = (\{v_0, v_1, v_2, v_3, v_4\}, \{e_0, e_1, e_2, \ldots, e_7\})$. We denote the number of vertices in a graph by $n = |V|$ and the number of edges by $|E|$. We can specify an edge in terms of the two edges it connects (these are called its *end-points*. If the end-points of some edge e are v_i and v_j, then we can write $e = (v_i, v_j)$ (as well as $e = (v_j, v_i)$ if e is undirected). So, we can define the graph in Fig. 2.9 as $G = (V, E), V = (\{v_0, v_1, v_2, v_3, v_4\})$, $E = (\{(v_0, v_0), (v_0, v_1), (v_1, v_4), (v_4, v_1), (v_1, v_3), (v_1, v_2), (v_2, v_3), (v_3, v_4)\})$.

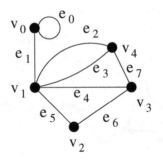

Fig. 2.9. Another graph

The *degree* of a vertex is the number of edges *incident* with it. In Fig. 2.9, the degree of vertex v_1 is 5, and that of v_2 is 2. Note several interesting properties of this graph. The first is that it has a *self-loop* – an edge (u, v) where $u = v$ (i.e., e_0). The second is that it contains *parallel edges* (i.e., e_2 and e_3). We will normally only be concerned with *simple* graphs (i.e., those without self-loops or parallel edges). A rather more interesting property of the graph is that it contains a *cycle* – a path through the graph that starts and ends at the *same* vertex. An example cycle on the graph depicted in Fig. 2.9 starts and ends at $v_1 - e_2 \to e_7 \to e_6 \to e_5$.

We are often interested in certain *other* properties of graphs. For example, we may want to ask the question "is there a path from u to v?", "is there a path containing each *edge* exactly once?", or "is there a cycle that visits each *vertex* exactly once?" Such questions concern the existence and/or construction of paths having particular properties.

Consider a salesperson who wishes to visit several cities connected by rail links. In order to save time and money, she would like to know if there exists an itinerary that visits every city precisely once. We model this situation by constructing a graph where vertices represent cities and edges the rail links. The problem, known as the *Hamiltonian Path Problem* [67], is very well-studied, and we will examine it in depth in subsequent chapters.

Another set of questions may try to *categorize* a graph. For example, two vertices v and w are said to be *connected* if there is a path connecting them. So a connected graph is one for which every pair of vertices has a path connecting them. If we asked the question "Is this graph *connected*?" of the graph depicted in Fig. 2.8, the answer would be "no", while the answer for the graph in Fig. 2.9 would be "yes".

The question of connectivity is important when considering, for example, telecommunications networks. These can be visualized as a graph, where vertices represent communication "stations", and edges represent the links between them. The removal of a vertex (and all edges incident to it) corresponds to the situation where a station malfunctions. By removing vertices from a graph and examining its subsequent connectivity, we can identify crucial stations and perhaps redesign the network such that their removal does not lead to the graph becoming disconnected (at the very least).

Another category of graphs is made up of those that can have their vertices "colored" using one of three colors (say, red, green, and blue). These graphs fall into the "3-vertex-colorable" category. Graphs that do and do not fall into this category are depicted in Fig. 2.10. The problem is to decide whether three colors are sufficient to achieve such a coloring for an *arbitrary* graph [67]. If we consider a graph with n vertices, there are clearly 3^n possible ways of assigning colors to vertices, but only a fraction of them will encode proper colorings.

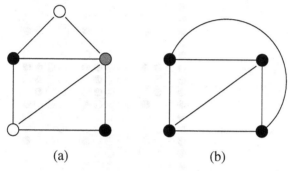

(a)   (b)

Fig. 2.10. (a) 3-colorable. (b) Non-3-colorable graph

In order to clarify this, consider Fig. 2.11a. All possible three-colorings of G (a) are depicted graphically in (b), with all proper colorings framed. A proper coloring is highlighted in (c), and an illegal coloring in (d) (note how v_2 and v_3 are colored the same).

Coloring could be applied to a situation where, for example, a company manufactures several chemicals, certain pairs of which could explode if brought into close contact. The company wishes to partition its warehouse into sealed compartments and store incompatible chemicals in different compartments. Obviously, the company wishes to minimize the amount of building work required, and so needs to know the least number of compartments into which the warehouse should be partitioned. We can construct a graph with vertices representing chemicals, and edges between vertices representing incompatibilities. The answer to the question of how many partitions are required is therefore equal to the smallest number of colors required to obtain a proper (i.e., legal) coloring of the graph.

Other questions involve asking whether or not a graph contains a *subgraph* with its own particular properties. For example, "Does it contain a cycle?" The cycle shown in the example graph G depicted in Fig. 2.9 is considered a subgraph of G, as we see in Fig. 2.12.

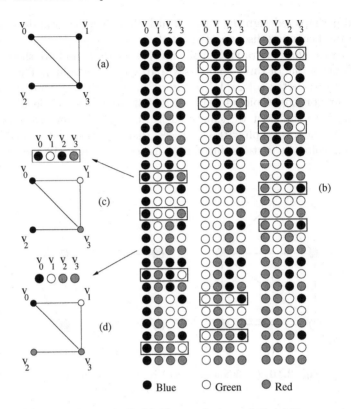

Fig. 2.11. (a) Example graph G. (b) Three-colorings of G. (c) Proper coloring. (d) Illegal coloring

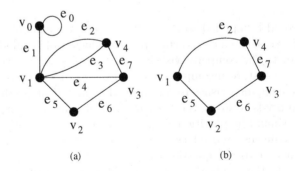

(a)   (b)

Fig. 2.12. (a) Example graph G. (b) Subgraph of G

Another problem may be concerned with *cliques*. A clique is a set of vertices in an undirected graph in which each vertex is connected to every other vertex. Cliques are *complete graphs* on n vertices, where n is the number of vertices in the clique. The *maximum clique* problem is concerned with finding the largest clique in a graph.

For example, consider the graph depicted in Fig. 2.13. It contains a subgraph that is a clique of size 4.

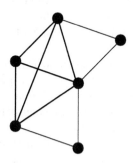

Fig. 2.13. Graph containing a clique (shown in bold)

The problem of identifying "clusters" of related objects is often equivalent to finding cliques in graphs. One "real world" application of this was developed by the US Internal Revenue Service to detect organized tax fraud, where groups of "phony" tax forms are submitted. A graph is constructed, where vertices correspond to tax return forms and edges link any forms that look similar. The existence of a large clique suggests large-scale organized fraud [146].

As we can see, our problem is to either *decide* if a graph has certain properties or *construct* from it a path or subgraph with certain properties. We show in a later chapter how this may be achieved in the context of models of DNA computation.

Another type of graph is the *tree*. These are connected, undirected, acyclic graphs. This structure is accessed at the *root* vertex (usually drawn, perhaps confusingly, at the top of any diagrammatic representation). Each vertex in a tree is either a *leaf* node or an *internal* node (the terms vertex and node are interchangeable). An internal node has one or more *child* nodes, and is called the *parent* of the child node. An example tree is depicted in Fig. 2.14.

2.6 Computational Complexity

It should be clear that different algorithms require different resources. A program to sort one hundred integers into ascending order will "obviously" run

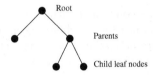

Fig. 2.14. An example tree

faster (on the same computer) than one used to calculate the first million prime numbers. However, it is important be able to establish this fact with certainty. The field of computational complexity is concerned with the resources required by algorithms.

Complexity may be defined as "the intrinsic minimum amount of resources, for instance, memory or time, needed to solve a problem or execute an algorithm." In practice, we generally consider *time* as the limiting factor (that is, when discussing the relative complexity of algorithms, we consider the time taken for them to achieve their task).

We measure the time complexity of an algorithm in terms of the number of computational "steps" it takes. For example, consider the factoring algorithm given in Sect. 2.4. We assume for the sake of argument than $n > 0$. Step 1 involves reading n. Step 2 is the conditional, checking if $n \leq 0$. Step 3 assigns the value 1 to the *power* variable. Then we enter a loop. This is the crucial part of the analysis, as it involves the section of the algorithm where most work is done. Step 4 checks the value of n. As long as $n > 0$, we execute two assignment steps (to *power* and n). We then execute a final step to print out the value of *power*. As *counter* has the value $n - 1$, this algorithm takes two steps if $n \leq 0$, and $5 + (n*3)$ steps if $n \geq 0$. Note that the constant 5 includes the final "while" step.

However, consider the implementation of this algorithm on a computer. The running time of the program, given a particular value of n, still depends on two factors:

1. The computer on which the program is run. Supercomputers execute instructions far more rapidly than personal computers,
2. The conversion of the pseudo-code into machine language. Programs to do this are called *compilers*, and some compilers generate more efficient machine code than others. The number of machine instructions used to implement a particular pseudo-code statement will vary from compiler to compiler.

We cannot therefore make statements like "this program will take 0.52 seconds to run for an input value of $n = 1000$", unless we know the precise details of the machine and the compiler. Even then, the fact that the machine may or may not be carrying out other time-consuming tasks renders such predictions meaningless. Nobody is interested in the time complexity of an algorithm compiled in a particular way running on a particular machine in a particular state. We therefore describe the running time of an algorithm using "big-oh" notation, which allows us to hide constant factors, such as the average number of machine instructions generated by a particular compiler and the average time taken for a particular machine to execute an instruction. So, rather than saying that the factoring algorithm takes $5 + (n * 3)$ time, we strip out the constants and say it takes $O(n)$[1] time.

The importance of this method of measuring complexity lies in determining whether of not an algorithm is suitable for a particular task. The fundamental question is whether or not an algorithm will be too slow given a big enough input, regardless of the efficiency of the implementation or the speed of the machine on which it will run.

Consider two algorithms for sorting numbers. The first, quicksort [48, page 145] runs, on average, in time $O(nlogn)$. The second, bubble sort [92], runs in $O(n^2)$. To sort a million numbers, quicksort takes, on average, 6,000,000 steps, while bubble sort takes 1,000,000,000,000. Consequently, quicksort running on a home computer will beat bubble sort running on a supercomputer!

An unfortunate tendency among those unfamiliar with computational complexity theory is to argue for "throwing" more computer power at a problem until it yields. Difficult problems can be cracked with a fast enough supercomputer, they reason. The only problems lie in waiting for technology to catch up or finding the money to pay for the upgrade. Unfortunately, this is far from the case.

Consider the problem of *satisfiability* (SAT) [48, page 996]. We formally define SAT in the next chapter, but for now consider the following problem. Given the circuit depicted in Fig. 2.15, is there a set of values for the variables (inputs) x and y that result in the circuit's output z having the value 1?

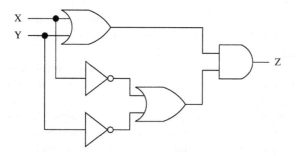

Fig. 2.15. Satisfiability circuit

---

[1] Read "big-oh of n", or "oh of n".

The function F computed by this circuit is (x OR y) AND (NOT x OR NOT y). In formal notation, this is expressed as $F = (x \vee y) \wedge (\overline{x} \vee \overline{y})$. OR is denoted by $\wedge$, AND by $\wedge$, and NOT by overlining the variable.

F is made up of two bracketed *clauses*, and the problem is to find values for the variables so that both clauses have the value 1, giving a value of 1 for F (observe that the values of the clauses are ANDed). This problem is known as the "satisfiability" problem because making all the clauses true is viewed as "satisfying" the clauses.

The most obvious method of solving this problem is to generate all possible choices for the variable values. That is, construct the truth table for the circuit. Table 2.5 shows a breakdown of the truth table for the circuit depicted in Fig. 2.15.

Table 2.5. Truth table for satisfiability circuit

x y	$(x \vee y)$	$(\overline{x} \vee \overline{y})$	z
0 0	0	1	0
0 1	1	1	1
1 0	1	1	1
1 1	1	0	0

Clearly, by inspection, only the combinations "x = 0, y = 1" and "x = 1, y = 0" are valid solutions to the problem. The current best algorithm essentially performs just this process of trying all possible 2^n choices for n variables. Therein lies the problem.

Consider a situation where the number of variables doubles to four. There are therefore $2^4 = 16$ combinations to check. Doubling the circuit size again means checking $2^8 = 256$ combinations. However, what if the function has 100 variables? The number of combinations to be checked is now a staggering 1,267,650,600,228,229,401,496,703,205,376!

This is what mathematicians refer to as a *combinatorial explosion*. This is a situation where work (or space required) increases by a factor of two, three, or more for each successive value of n.

Table 2.6 shows the time required by different algorithms with different time complexities for different values of n. We assume the algorithms are run by a machine capable of executing a million steps a second.

Clearly, algorithms requiring n (*linear*) or n^2 (*quadratic*) time may be feasible for relatively large problem instances, while algorithms requiring 2^n (*exponential*) time are clearly impractical for even small values of n.

Table 2.6. Running times of different algorithms for different values of n

n	$O(n)$	$O(n^2)$	$O(2^n)$	$O(n^n)$
1	0.000001 seconds	0.000001 seconds	0.000002 seconds	0.000001 seconds
5	0.000002 seconds	0.000025 seconds	0.000032 seconds	0.003125 seconds
10	0.00001 seconds	0.0001 seconds	0.001024 seconds	2.778 hours
50	0.00005 seconds	0.0025 seconds	35.7 years	2.87×10^{70} years

2.7 P and NP

We can therefore call an algorithm "fast" if the number of steps to solve a problem of size n is (no more than) some *polynomial* involving n. We define the *complexity class P* to mean the set of all *problems* (not algorithms) that have polynomial-time solutions. Therefore, the problem of sorting numbers is in P, since some solution (e.g., bubble sort) runs in $O(n^2)$ time, and n^2 is a polynomial.

By the late 1960s it became apparent that there were several seemingly simple problems for which no fast algorithms could be found, despite the best efforts of the algorithms community. In an attempt to classify *this* set of problems, Cook observed that in order for a problem to be solved in polynomial time one should be able (at the very least) to *verify* a given correct solution in polynomial time [47]. This observation holds because if we have a polynomial-time algorithm for a problem and someone gives us a *proposed* solution, we can always re-run the algorithm to obtain the correct solution and then compare the two, in polynomial time.

This led to the creation of the complexity class NP containing decision problems for which one can verify the solution in polynomial time. Cook also showed that within NP lies a set of problems that are the hardest of them all. If a polynomial-time algorithm exists for any one of these problems then *all* NP problems can be solved in polynomial time. This fact is known as *Cook's Theorem* [47], and is one of the most profound results in theoretical computer science. The class of those "hardest" problems in NP is known as *NP-complete* problems, of which satisfiability is the archetype.

Soon after Cook's paper, Karp [89] proved that several interesting problems could also be shown to be NP-complete. The list of NP-complete problems has grown considerably since, and the standard text may be found at [63]. The list includes many problems of great theoretical and practical significance, such as network design, scheduling, and data storage.

2.8 Summary

In this chapter we provided an introduction to the theory of computer science. We described the fundamental "building blocks" of computers, *logic gates,* and showed how they may be pieced together to perform computations. We then considered the nature of *computation* itself, and introduced the concept of an *algorithm*. This motivated the study of machine models, or *models of computation*. We introduced several such models (the *finite-state automaton*, the *Turing machine*, and the *Random Access Machine*), and described their features, strengths, and weaknesses. We then considered the implementation of algorithms within these models, introducing the organization of information into *data structures*. We examined a simple data structure, the *array*, before considering a more complex structure, the *graph*. We then highlighted the fact that different algorithms require different resources, introducing the key concept of *computational complexity*. We described the importance of this idea, and showed how to calculate the complexity of algorithms. This then motivated a discussion of how to choose an algorithm for a particular problem. We concluded with a description of *complexity classes*, and introduced the *NP-complete problems*, for which no fast algorithms yet exist.

2.9 Bibliographical Notes

A standard textbook on computer architecture is [149], which includes a treatment of Boolean algebra. Algorithms and complexity are introduced in [48] (which also covers sorting networks, encountered in Chap. 5). Another classic text on data structures and algorithms is [5]. For an excellent introduction to the theory of graphs, see [67]. The controversial[2] paper that first established that any planar graph is four colorable is published as both a book [16] and a paper [15]. Finally, a discussion of why some problems are inherently difficult and a treatment of state-of-the art solution methods are given in [107].

---

[2] Due to its reliance on computer-assisted proof.

3

Models of Molecular Computation

"... there's a statue inside every block of stone." – George Orwell, Coming up for Air

3.1 Introduction

The purpose of this chapter is to describe several examples of the various models of molecular computation that have been proposed in the literature. Note that we have used the term "molecular" rather than the more specific "DNA", as we wish to abstract away from the (perhaps) intended biological substrate for each model, and concentrate on the *computational* features of the machine model.

We may describe abstract models of computation without necessarily considering their implementation. In [64], for example, for the sake of emphasising what is inherently parallelisable within problems, the authors disregard constraints of implementation. However, the operation sets within the models described here are constrained by the availability of various molecular manipulation techniques. The implementation of abstract operations will largely determine the success or failure of a model. Many of the models described in this chapter use abstract operations common to the others, such as set union. However, even though models may utilize similar operations (e.g., removal of an element from a set), the chosen implementation method may differ from model to model. Details may impact implementation in various ways:

1. The volume of DNA required (analogous to *space* in complexity theoretical terms) to perform the computation may vary by exponential factors.
2. Each operation takes a certain amount of time to implement in the laboratory, and so the sequence of operations performed determines the overall time complexity of the algorithm. Thus, the techniques chosen have a direct bearing on the efficiency of a DNA-based algorithm. In addition,

the time taken to construct the initial set of strings and read out the final solution may be very time consuming, and must also be taken into account.

3. Each laboratory technique has associated with it a nonzero error rate. Some techniques are far more error-prone than others, so the choice of laboratory techniques directly affects the probability of success of a DNA-based algorithm.

We therefore focus on *abstract models* in this chapter, and consider their *physical* implementation in Chap. 5. The models fall into four natural categories:

- Filtering
- Splicing
- Constructive
- Membrane

3.2 Filtering Models

In all filtering models (motivated by Adleman [3] and contemporaneously generalized by Lipton [98] and Amos et al. [94, 12]), a computation consists of a sequence of operations on finite *multi-sets* of strings. Multi-sets are sets that may contain more than one copy of the same element. It is normally the case that a computation begins and terminates with a single multi-set. Within the computation, by applying legal operations of a model, several multi-sets may exist at the same time. We define operations on multi-sets shortly, but first consider the nature of an *initial set*.

An initial multi-set consists of strings which are typically of length $O(n)$ where n is the problem size. As a subset, the initial multi-set should include all possible solutions (each encoded by a string) to the problem to be solved. The point here is that the superset, in any implementation of the model, is supposed to be relatively easy to generate as a starting point for a computation. The computation then proceeds by *filtering out* strings which cannot be a solution. For example, the computation may begin with a multi-set containing strings representing all *possible* three-colorings of a graph, and then proceed by removing those that encode *illegal* colorings.

To give another example, if the problem is to generate a permutation of the integers $1, \ldots, n$, then the initial multi-set might include all strings of the form $p_1 i_1 p_2 i_2 \ldots p_n i_n$ where each i_k may be any of the integers in the range $[1, \ldots, n]$ and p_k encodes the information "position k." Here, as will be typical for many computations, the multi-set has cardinality which is exponential in the problem size. For our example of finding a permutation, we should filter out all strings in which the same integer appears in at least two locations p_k. Any of the remaining strings is then a legal solution to the problem.

We now describe the important features of the various filtering models.

Unrestricted model

Adleman [3] provided the impetus for recent work through his experimental solution to the Hamiltonian Path Problem. This solution, however, was not expressed within a formal model of computation, and is therefore described later in Chap. 5. In [98], Lipton considered Adleman's specific model and showed how it can encompass solutions to one other NP-complete problem. Here we summarize the operations within Adleman's subsequent *unrestricted* model [4]. All operations are performed on sets of strings over some alphabet α.

- *separate*(T, S). Given a set T and a substring S, create two new sets $+(T, S)$ and $-(T, S)$, where $+(T, S)$ is all strings in T containing S, and $-(T, S)$ is all strings in T *not* containing S.
- *merge*$(T_1, T_2, \ldots, T_n)$. Given set $T_1, T_2, \ldots, T_n$, create $\cup(T_1, T_2, \ldots, T_n) = T_1 \cup T_2 \cup \ldots T_n$.
- *detect*(T). Given a set T, return *true* if T is nonempty, otherwise return *false*.

For example, given $\alpha = \{A, B, C\}$, the following algorithm returns *true* only if the initial multi-set contains a string composed entirely of "A"s:

    Input(T)
    T ← −(T, B)
    T ← −(T, C)
    Output(detect(T))

In [4] Adleman describes an algorithm for the *3-vertex-colorability* problem. Recall, from Chap. 2, that in order to obtain a proper coloring of a graph $G = (V, E)$ colors are assigned to the vertices in such a way that no two adjacent vertices are similarly colored.

We now describe Adleman's algorithm in detail. The initial set, T, consists of strings of the form $c_1, c_2, \ldots, c_n$, where $c_i \in \{r_i, g_i, b_i\}$ and n is $|V|$, the number of vertices in G. Thus each string represents one possible (not necessarily proper) coloring of the given graph. With reference to Fig. 2.11, the coloring represented in (c) would be encoded by the string b_0, g_1, b_2, r_3, and the coloring in (d) would be encoded by b_0, g_1, r_2, r_3.

We assume that all possible colorings are represented in T. The algorithm proceeds as follows:

(1) read initial set T
(2) for each vertex do
(3)   From T, create *red* tube containing strings encoding this
        vertex red, and create *blue/green* tube containing
        all other strings
(4)   Create *blue* tube from *blue/green* tube,
        and create *red* tube from remaining strings

(5)   for all vertices adjacent to this vertex do
(6)   From *red*, remove strings encoding
                   adjacent vertex red
(7)   From *blue*, remove strings encoding
                   adjacent vertex blue
(8)   From *green*, remove strings encoding
                   adjacent vertex green
(9)   **end for**
(10)   Merge red, green and blue tubes to form *new* tube T
(11) **end for**
(12) Read what is left in T

Or more formally:

(1) Input(T)
(2) **for** $i = 1$ to n **do begin**
(3) $T_r \leftarrow +(T, r_i)$ **and** $T_{bg} \leftarrow -(T, r_i)$
(4) $T_b \leftarrow +(T_{bg}, b_i)$ **and** $T_g \leftarrow -(T_{bg}, b_i)$
(5)   **for all** j such that $< i, j >\in E$ **do begin**
(6) $T_r \leftarrow -(T_r, r_j)$
(7) $T_g \leftarrow -(T_g, g_j)$
(8) $T_b \leftarrow -(T_b, b_j)$
(9)   **end for**
(10) $T \leftarrow merge(T_r, T_g, T_b)$
(11) **end for**
(12) Output(detect(T))

At Step 1 we input all possible colorings of the graph. Then, for each vertex $v_i \in V$ we perform the following steps: split T into three sets, T_r, T_g, T_b, where T_r contains only strings containing r_i, T_g contains only strings containing g_i, and T_b contains only strings containing b_i (Steps 3–4). Then, for each edge $< i, j >\in E$, we remove from these sets any strings containing $c_i = c_j$ (i.e., those strings encoding colorings where adjacent vertices i and j are colored the same) (Steps 5–9). Then, these sets are merged, forming the new set T (Step 10), and the algorithm proceeds to the next vertex (Step 11). After the coloring constraints for each vertex have been satisfied, we perform a detection (Step 12). If T is nonempty then any string in T encodes a proper 3-vertex-coloring of G.

Satisfiability model

Lipton [98] described a solution to another NP-complete problem, namely the so-called *satisfiability* problem (SAT). SAT may be phrased as follows: given a finite set $V = \{v_1, v_2, \ldots, v_n\}$ of logical variables, we define a *literal* to be a variable, v_i, or its complement, $\overline{v_i}$. If v_i is *true* then $\overline{v_i}$ is *false*, and vice-versa. We define a *clause*, C_j, to be a set of literals $\{v_1^j, v_2^j, \ldots, v_l^j\}$. An instance, I, of SAT consists of a set of clauses. The problem is to assign a Boolean value to each variable in V such that at least one variable in each clause has the value *true*. If this is the case we may say that I has been *satisfied*.

Although Lipton does not explicitly define his operation set in [98], his solution may be phrased in terms of the operations described by Adleman in [4]. Lipton employs the *merge, separate*, and *detect* operations described above. The initial set T contains many strings, each encoding a single n-bit sequence. All possible n-bit sequences are represented in T. The algorithm proceeds as follows:

(1) Create initial set, T
(2) For each clause do begin
(3)   For each literal v_i do begin
(4)   if $v_i = x_j$ extract from T strings encoding $v_i = 1$ else
                 extract from T strings encoding $v_i = 0$
(5)   End for
(6)   Create new set T by merging extracted strings
(7) End for
(8) If T nonempty then I is satisfiable

The pseudo-code algorithm may be expressed more formally thus:

(1) Input(T)
(2) **for** $a = 1$ to $|I|$ **do begin**
(3)   **for** $b = 1$ to $|C_a|$ **do begin**
(4)   **if** $v_b^a = x_j$ **then** $T_b \leftarrow +(T, v_b^a = 1)$
                 else $T_b \leftarrow +(T, v_b^a = 0)$
(5)   **end for**
(6) $T \leftarrow merge(T_1, T_2, \ldots, T_b)$
(7) **end for**
(8) Output(detect(T))

Step 1 generates all possible n-bit strings. Then, for each clause $C_a = \{v_1^a, v_2^a, \ldots, v_l^a\}$ (Step 2) we perform the following steps. For each literal v_b^a (Step 3) we operate as follows: if v_b^a computes the positive form then we extract from T all strings encoding 1 at position v_b^a, placing these strings in T_b; if v_b^a computes the negative form we extract from T all strings encoding 0 at position v_b^a, placing these strings in T_b (Step 4); after l iterations, we have satisfied every variable in clause C_a; we then create a new set T from the union of sets $T_1, T_2, \ldots, T_b$ (Step 6) and repeat these steps for clause $C_a + 1$ (Step 7). If any strings remain in T after all clauses have been operated upon, then I is satisfiable (Step 8).

Parallel filtering model

A detailed description of our parallel filtering model appears in [12]. This model was the first to provide a formal framework for the easy description of DNA algorithms for *any* problem in the complexity class NP. Lipton claims some generalisation of Adleman's style of computation in [98], but it is difficult to see how algorithms for different problems may be *elegantly* and *universally* expressed within his model. Lipton effectively uses the same operations as Adleman, but does not explicitly describe the operation set. In addition, he describes only one algorithm (3SAT), whereas in subsequent sections we show how our model provides a natural description for any NP-complete problem through many examples.

As stated earlier, within our model all computations start with the construction of the initial set of strings. Here we define the basic legal operations on sets within the model. Our choice is determined by what we know can be effectively implemented by very precise and complete chemical reactions within the DNA implementation. The operation set defined here provides the power we claim for the model but, of course, it might be augmented by additional operations in the future to allow greater conciseness of computation. The main difference between the parallel filtering model and those previously proposed lies in the implementation of the removal of strings. All other models propose *separation* steps, where strings are *conserved*, and may be used later in the computation. Within the parallel filtering model, however, strings that are removed are *discarded*, and play no further part in the computation. This model is the first exemplar of the so-called "mark and destroy" paradigm of molecular computing.

- $remove(U, \{S_i\})$. This operation removes from the set U, in parallel, any string which contains at least one occurrence of any of the substrings S_i.
- $union(\{U_i\}, U)$. This operation, in parallel, creates the set U which is the set union of the sets U_i.
- $copy(U, \{U_i\})$. In parallel, this operation produces a number of copies, U_i, of the set U.
- $select(U)$. This operation selects an element of U at random; if U is the empty set then *empty* is returned.

From the point of view of establishing the parallel time complexities of algorithms within the model, these basic set operations will be assumed to take *constant time*. However, this assumption is reevaluated in Chap. 4.

A first algorithm

We now provide our first algorithmic description within the model. The problem solved is that of generating the set of all permutations of the integers 1 to n. A permutation is a *rearrangement* of a set of elements, where none are removed, added, or changed. The initial set and the filtering out of strings which are not permutations were essentially described earlier. Although not NP-complete, the problem does of course have exponential input and output.

The algorithmic description below introduces a format that we utilize elsewhere. The particular device of copying a set (as in copy($U, \{U_1, U_2, \ldots, U_n\}$)) followed by parallel *remove* operations (as in remove($U_i, \{p_j \neq i, p_k i\}$)) is a very useful compound operation, as we shall later see in several algorithmic descriptions. Indeed, it is precisely this use of *parallel filtering* that is at the core of most algorithms within the model.

Problem: Permutations
Generate the set P_n of all permutations of the integers $\{1, 2, \ldots, n\}$.

Solution:

- Input: The input set U consists of all strings of the form $p_1 i_1 p_2 i_2 \ldots p_n i_n$ where, for all j, p_j uniquely encodes "position j" and each i_j is in $\{1, 2, \ldots, n\}$. Thus each string consists of n integers with (possibly) many occurrences of the same integer.

- Algorithm

  for $j = 1$ to $n - 1$ **do**
      begin
      copy($U, \{U_1, U_2, \ldots, U_n\}$)
      for $i = 1, 2, \ldots, n$ and all $k > j$
            in parallel do remove($U_i, \{p_j \neq i, p_k i\}$)
      union($\{U_1, U_2, \ldots, U_n\}, U$)
      end
  $P_n \leftarrow U$

- Complexity: $O(n)$ parallel time.

After the jth iteration of the **for** loop, the computation ensures that in the surviving strings the integer i_j is not duplicated at positions $k > j$ in the string. The integer i_j may be any in the set $\{1, 2, \ldots, n\}$ (which one it is depends on which of the sets U_i the containing string survived). At the end of the computation each of the surviving strings contains exactly one occurrence of each integer in the set $\{1, 2, \ldots, n\}$ and so represents one of the possible permutations. Given the specified input, it is easy to see that P_n will be the set of all permutations of the first n natural numbers. As we shall see, production of the set P_n can be a useful sub-procedure for other computations.

Algorithms for a selection of NP-complete problems

We now describe a number of algorithms for graph-theoretic NP-complete problems (see [67], for example). Problems in the complexity class NP seem to have a natural expression and ease of solution within the model. We describe linear-time solutions although, of course, there is frequently an implication of an exponential number of processors available to execute any of the basic operations in unit time.

The 3-vertex-colorability problem

Problem: Three coloring
Given a graph $G = (V, E)$, find a 3-vertex-coloring if one exists, otherwise return the value *empty*.

Solution:

- Input: The input set U consists of all strings of the form $p_1 c_1 p_2 c_2 \ldots p_n c_n$ where $n = |V|$ is the number of vertices in the graph. Here, for all i, p_i uniquely encodes "position i" and each c_i is any one of the "colors" 1, 2, or 3. Each such string represents one possible assignment of colors to the vertices of the graph in which, for each i, color c_i is assigned to vertex i.

- Algorithm:

  for $j = 1$ to n **do**
  　　begin
  　　copy$(U, \{U_1, U_2, U_3\})$
  　　for $i = 1$, 2 and 3, and all k such that $(j, k) \in E$
  　　　　in parallel do remove$(U_i, \{p_j \neq i, p_k i\})$
  　　union$(\{U_1, U_2, U_3\}, U)$
  　　end
  　select(U)

- Complexity: $O(n)$ parallel time.

After the jth iteration of the **for** loop, the computation ensures that in the remaining strings vertex j (although it may be colored 1, 2, or 3 depending on which of the sets U_i it survived in) has no adjacent vertices that are similarly colored. Thus, when the algorithm terminates, U only encodes legal colorings if any exist. Indeed, every legal coloring will be represented in U.

The Hamiltonian path problem

A Hamiltonian path between any two vertices u, v of a graph is a path that passes through every vertex in $V - \{u, v\}$ precisely once [67].

Problem: Hamiltonian path
Given a graph $G = (V, E)$ with n vertices, determine whether G contains a Hamiltonian path.

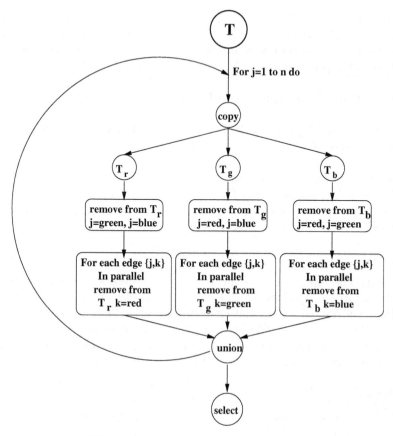

Fig. 3.1. 3-coloring algorithm flowchart

Solution:

- Input: The input set U is the set P_n of all permutations of the integers from 1 to n as output from **Problem: Permutations**. An integer i at position p_k in such a permutation is interpreted as follows: the string represents a candidate solution to the problem in which vertex i is visited at step k.

- Algorithm:

  for $2 \leq i \leq n-1$ and j, k such that $(j, k) \notin E$
     in parallel do remove $(U, \{jp_ik\})$
  select(U)

- Complexity: Constant parallel time given P_n.

In surviving strings there is an edge of the graph for each consecutive pair
of vertices in the string. Since the string is also a permutation of the vertex
set it must also be a Hamiltonian path. Of course, U will contain every legal
solution to the problem.

The subgraph isomorphism problem

Given two graphs G_1 and G_2 the following algorithm determines whether G_2
is a subgraph of G_1.

Problem: Subgraph isomorphism
Is $G_2 = (V_2, E_2)$ a subgraph of $G_1 = (V_1, E_1)$? By $\{v_1, v_2, \ldots, v_s\}$ we denote
the vertex set of G_1; similarly the vertex set of G_2 is $\{u_1, u_2, \ldots, u_t\}$ where,
without loss of generality, we take $t \leq s$.

Solution:

- Input: The input set U is the set P_s of permutations output from the
  Permutations algorithm. For $1 \leq j \leq t$ an element $p_1 i_1 p_2 i_2 \ldots p_s i_s$ of
  P_s is interpreted as associating vertex $p_j \in \{u_1, u_2, \ldots, u_t\}$ with vertex
  $i_j \in \{v_1, v_2, \ldots, v_s\}$. The algorithm is designed to remove any element
  which maps vertices in V_1 to vertices in V_2 in a way which does not reflect
  the requirement that if $(p_s, p_t) \in E_1$ then $(i_s, i_t) \in E_2$.

- Algorithm:

  for $j = 1$ to $t - 1$ **do**
      begin
      $\text{copy}(U, \{U_1, U_2, \ldots, U_t\})$
      for all $l, j < l \leq t$ such that $(p_j, p_l) \in E_2$ and $(ij, i_l) \notin E_1$
                  in parallel do $\text{remove}(U_j, \{p_l i_l\})$
      $\text{union}(\{U_1, U_2, \ldots, U_t\}, U)$
      end
  $\text{select}(U)$

- Complexity: $\mathrm{O}(|V_s|)$ parallel time.

For any remaining strings, the first t pairs $p_l i_l$ represent a one-to-one association of the vertices of G_1, with the vertices of G_2 indicating the subgraph of
G_1 which is isomorphic to G_2. If $\text{select}(U)$ returns the value *empty* then G_2
is not a subgraph of G_1.

The maximum clique problem
A clique K_i is the complete graph on i vertices [67]. The problem of finding a maximum independent set is closely related to the maximum clique problem.

Problem: Maximum clique
Given a graph $G = (V, E)$ determine the largest i such that K_i is a subgraph of G. Here K_i is the complete graph on i vertices.

Solution:

- In parallel run the subgraph isomorphism algorithm for pairs of graphs (G, K_i) for $2 \leq i \leq n$. The largest value of i for which a nonempty result is obtained solves the problem.
- Complexity: $O(|V|)$ parallel time.

The above examples fully illustrate the way in which the NP-complete problems have a natural mode of expression within the model. The mode of solution fully emulates the definition of membership of NP: that instances of problems have candidate solutions that are polynomial-time verifiable and that there are generally an exponential number of candidates.

Sticker model

We now introduce an alternative filtering-style model due to Roweis et al. [133], named the *sticker model*. Within this model, operations are performed on multisets of strings over the binary alphabet $\{0, 1\}$. *Memory* strands are n characters in length, and contain k nonoverlapping substrings of length m. Substrands are numbered contiguously, and there are no gaps between them (Fig. 5.10a). Each substrand corresponds to a Boolean variable (or *bit*), so within the model each substrand is either *on* or *off*. We use the terms *bit* and *substrand* interchangeably

We now describe the operations available within the sticker model. They are very similar in nature to those operations already described previously, and we retain the same general notation. A *tube* is a multiset, its members being memory strings.

- *merge*. Create the multiset union of two tubes.
- *separate*(N, i). Given a tube N and an integer i, create two new tubes $+(N, i)$ and $-(N, i)$, where $+(N, i)$ contains all strings in N with substrand i set to *on*, and $-(N, i)$ contains all strings in N with substrand i set to *off*.
- *set*(N, i). Given a tube N and an integer i, produce a new tube *set*(N, i) in which the ith substrand of every memory strand is turned *on*.
- *clear*(N, i). Given a tube N and an integer i, produce a new tube *set*(N, i) in which the ith substrand of every memory strand is turned *off*.

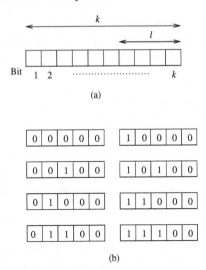

(a)

(b)

Fig. 3.2. (a) General memory strand scheme. (b) Complete (5,2) library

As in the models described previously, computations within the model consist of sequences of operations taken from the available set. The final result is read from the output tube by determining the contents of the memory strands it contains; if the output tube is empty then this fact is reported.

The initial input to a computation consists of a *library* of memory strands. A (k, l) library, where $1 \leq l \leq k$, consists of memory strands with k substrands, the first l of which may be on or off, and the last $k - l$ of which are off. Therefore, in an initial tube the first l substrands represent the *input* to the computation, and the remaining $k - l$ substrands are used for working storage and representation of output. For example, a complete $(5,2)$ library is depicted in Fig. 3.2b.

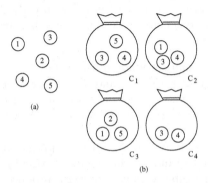

Fig. 3.3. (a) Five possible objects. (b) Four different bags

We now illustrate a computation within the model. Again, the general strategy adopted here is to generate all possible solutions to a given problem and then exhaustively filter until either a solution is found or it can be stated that no solution exists.

Following [133] (and the description in [120]), we describe a sticker model solution to the *Minimal Set Cover* problem. This may be stated informally thus; given a set of different objects and a set of bags containing various subsets of the possible set of objects, what is the smallest number of bags a person must hold to ensure that he or she holds at least one of each object?

This is illustrated in Fig. 3.3, where (a) represents a set of five different numbered objects and (b) four different bags, each containing different subsets of the possible set.

Formally, given a finite size $S = \{2, \ldots, p\}$ (the objects), and a finite collection $\{C_1, \ldots, C_q\}$ of subsets of S, find the smallest subset I of $\{1, 2, \ldots, q\}$ such that $\bigcup_{i \in I} C_i = S$.

We now describe the sticker model solution to the problem instance depicted in Fig. 3.3 ($p = 5, q = 4$). Memory strands have $k = p + q$ substrands, and the initial tube N_0 contains a $(p + q, q)$ library. Each memory strand represents a possible subset of bags taken from the set C, with the first q substrands encoding the presence or absence of bags. If substrand i is set to on, then bag C_i is present in the subset; if substrand i is set to off, C_i is not present. For example, a memory strand with its first four substrands set to 1011 would encode a subset containing C_1, C_3 and C_4, while one with its first four substrands set to 0010 would encode a subset containing only C_3. The working substrands represent the subset of objects encompassed by the bags "held." For example, with reference to Fig. 3.3, a memory strand encoding "010110110" is interpreted as follows; the first four substrands encode the fact that bags C_2 and C_4 are held, and $C_2 \cup C_4 = \{1, 3, 4\}$, a fact that is encoded in the last five substrands.

Of course, when the initial library is created the working substrands are all set to off, so we must first set them appropriate values. This is achieved as follows: for each strand M, look at the first q substrands; for every i^{th} substrand that is turned on, use the *set* operation to turn on those among the last p working substrands that represent elements of C_i. Simply phrased, for a given memory strand, take the total subset of objects encoded by all of the bags it holds, and turn on the substrands representing each object from that subset. We represent the contents of N_0 after setup in Table 3.1. Note that *on* is represented by 1, and *off* by 0.

Once we have this set of strands, we wish to retain only those that encode a subset of bags that *cover* the set S (that is, subsets that contain one of each object in the full set). We achieve this by discarding those strands that do *not* have each of their five working substrands (representing the objects) set to on (or 1). This yields the strands depicted in Table 3.2.

Once we have established a set of coverings, the next task is to find in that a covering that uses the smallest number of bags (if one exists). This

Table 3.1. N_0 after setup

M	C_1	C_2	C_3	C_4	1	2	3	4	5	Subset
1	0	0	0	0	0	0	0	0	0	empty
2	0	0	0	1	0	0	1	1	5	(3,4,5)
3	0	0	1	0	1	1	0	0	1	(1,2,5)
4	0	0	1	1	1	1	1	1	1	(1,2,3,4,5)
5	0	1	0	0	1	0	1	1	0	(1,3,4)
6	0	1	0	1	1	0	1	1	0	(1,3,4)
7	0	1	1	0	1	1	1	1	1	(1,2,3,4,5)
8	0	1	1	1	1	1	1	1	1	(1,2,3,4,5)
9	1	0	0	0	0	0	1	1	1	(3,4,5)
10	1	0	0	1	0	0	1	1	1	(3,4,5)
11	1	0	1	0	1	1	1	1	1	(1,2,3,4,5)
12	1	0	1	1	1	1	1	1	1	(1,2,3,4,5)
13	1	1	0	0	1	0	1	1	1	(1,3,4,5)
14	1	1	0	1	1	0	1	1	1	(1,3,4,5)
15	1	1	1	0	1	1	1	1	1	(1,2,3,4,5)
16	1	1	1	1	1	1	1	1	1	(1,2,3,4,5)

Table 3.2. N_0 after culling

M	C_1	C_2	C_3	C_4	1	2	3	4	5	Subset
4	0	0	1	1	1	1	1	1	1	(1,2,3,4,5)
7	0	1	1	0	1	1	1	1	1	(1,2,3,4,5)
8	0	1	1	1	1	1	1	1	1	(1,2,3,4,5)
11	1	0	1	0	1	1	1	1	1	(1,2,3,4,5)
12	1	0	1	1	1	1	1	1	1	(1,2,3,4,5)
15	1	1	1	0	1	1	1	1	1	(1,2,3,4,5)
16	1	1	1	1	1	1	1	1	1	(1,2,3,4,5)

is achieved by sorting tube N_0 into a number of tubes $N_0, N_1, \ldots, N_q$, where tube N_i contains memory strands encoding coverings using i bags. This sorting process may be visualized as follows: imagine a row of tubes, stretching left to right, with N_0 at the left, and N_q at the right. We loop, maintaining a counter i from 1 to 4, each time dragging right one tube any strands containing bag i. This process is visualized in Fig. 3.4. Note, for example, how strand (3,4) is left in N_0 until the counter reaches 3, at which point it is dragged right into N_1. In this way, we end the computation with a set of tubes where $N_i, i \geq 1$ contains only strands encoding coverings using i bags. This process may be thought of as an inverted electronic version of gel electrophoresis (see Chap. 1), where "heavier" strands (those featuring more bags) are dragged further to the right than "lighter" strands.

We then start with the smallest indexed tube (N_0), and read its contents (if any). If that tube is empty, we move on to the next highest indexed tube

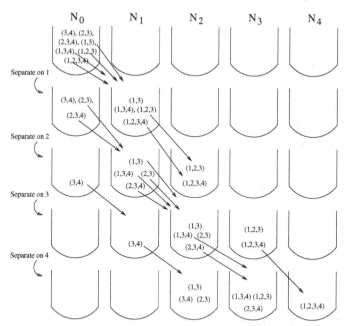

Fig. 3.4. Sorting procedure

and so on until we find a tube that contains a covering. In this case, tube N_2 contains three coverings, each using two bags. The algorithm is formally expressed within the sticker model as follows.

(1) Initialize (p,q) library in tube N_0
(2) **for** $i = 1$ to q **do begin**
(3) $N_0 \leftarrow separatei(+(N_0, i), -(N_0, i))$
(4)   **for** $j = 1$ to $| C_i |$
(5) $set(+(N_0, i), q + c_i^j)$
(6)   **end for**
(7) $N_0 \leftarrow merge(+(N_0, i), -(N_0, i))$
(8) **end for**

This section sets the object identifying substrands. Note that c_i^j denotes the jth element of set C_i. We now separate out for further use only those memory complexes where each of the last p substrands is set to on.

(1) **for** $i = q + 1$ to $q + p$ **do begin**
(2) $N_0 \leftarrow +(N_0, i)$
(3) **end for**

We now sort the remaining strands according to how many bags they encode.

  (1) **for** $i = 0$ to $q - 1$ **do begin**
  (2)   **for** $j - 1$ **down to** 0 **do begin**
  (3) $separate(+(N_j, i + 1), -(N_j, i + 1))$
  (4) $N_{j+1} \leftarrow merge(+(N_j, i + 1), N_{j+1})$
  (5) $N_j \leftarrow -(N_j, i + 1)$
  (6)   **end for**
  (7) **end for**

Line 3 separates each tube according to the value of i, and line 4 performs the right shift of selected strands. We then search for a final output:

  (1) Read N_1
  (2) **else if** empty **then** read N_2
  (3) **else if** empty **then** read N_3
  (4) ...

3.3 Splicing Models

Since any instance of any problem in the complexity class NP may be expressed in terms of an instance of any NP-complete problem, it follows that the multi-set operations described earlier at least implicitly provide sufficient computational power to solve any problem in NP. We do not believe they provide the full algorithmic computational power of a Turing Machine. Without the availability of string editing operations, it is difficult to see how the transition from one state of the Turing Machine to another may be achieved using DNA. However, as several authors have recently described, one further operation, the so-called *splicing* operation, will provide full Turing computability. Here we provide an overview of the various splicing models proposed.

Let S and T be two strings over the alphabet α. Then the *splice* operation consists of cutting S and T at specific positions and concatenating the resulting prefix of S with the suffix of T and concatenating the prefix of T with the suffix of S (Fig. 3.5). This operation is similar to the *crossover operation* employed by genetic algorithms [68, 96].

Splicing systems date back to 1987, with the publication of Tom Head's seminal paper [77] (see also [78]). In [50], the authors show that the generative power of *finite extended splicing systems* is equal to that of Turing Machines. For an excellent review of splicing systems, the reader is directed to [120].

Reif's PAM model

In [128], Reif within his so-called *Parallel Associative Memory Model* describes a *Parallel Associative Matching* (PA-Match) operation. The essential con-

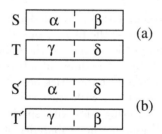

Fig. 3.5. (a) Two strings, S and T. (b) The result of $splice(S, T)$

stituent of the PA-Match operation is a restricted form of the splicing operation which we denote here by *Rsplice*, and describe as follows. If $S = S_1 S_2$ and $T = T_1 T_2$, then the result of $Rsplice(S, T)$ is the string $S_1 T_2$ *provided* $S_2 = T_1$, *but* has no value if $S_2 \neq T_1$.

Leading results of Reif [128], made possible by his PA-Match operation, concern the simulation of nondeterministic Turing Machines and the simulation of Parallel Random Access Machines. We can capture the spirit of his Turing Machine simulation through the *Rsplice* operation as follows: the initial tube in the simulation consists of all strings of the form $S_i S_j$ where S_i and S_j are encodings of *configurations* of the simulated nondeterministic Turing Machines, and such that S_j follows from S_i after one (of possibly many) machine cycle. By a configuration we mean here an instantaneous description of the Turing Machine capturing the contents of the tape, the machine state, and which tape square is being scanned. If the *Rsplice* operation is now performed between all pairs of initial strings, the tube will contain strings $S_k S_l$ where S_l follows from S_k after two machine cycles. Similarly, after t repetitions of this operation the tube will contain strings $S_m S_n$ where S_n follows from S_m after 2^t machine cycles. Clearly, if the simulated Turing Machine runs in time T, then after $O(\log T)$ operations the simulation will produce a tube containing strings $S_o S_f$ where S_o encodes the initial configuration and S_f encodes a final configuration.

3.4 Constructive Models

In this section we describe a constructive model of DNA computation, based on the principle of *self-assembly*. Molecular self-assembly gives rise to a vast number of complexes, including crystals (such as diamond) and the DNA double helix itself. It seems as if the growth of such structures are controlled, at a fundamental level, by natural computational processes. A large body of work deals with the study of self-organizational principles in natural systems, and the reader is directed to [40] for a comprehensive review of these.

We concentrate here on the *Tile Assembly Model*, due to Rothemund and Winfree [132]. This is a formal model for the self-assembly of complexes such as proteins or DNA on a square lattice. The model extends the theory of tiling by Wang tiles [152] to encompass the physics of self-assembly.

Within the model, computations occur by the self-assembly of square tiles, each side of which may labelled. The different labels represent ways in which tiles may bind together, the strength (or "stickiness") of the binding depending on the *binding strength* associated with each *side*. Rules within the system are therefore encoded by selecting tiles with specific combinations of labels and binding strengths. We assume the availability of an unlimited number of each tile. The computation begins with a specific seed tile, and proceeds by the addition of single tiles. Tiles bind together to form a growing complex representing the state of the computation only if their binding interactions are of sufficient strength (i.e., if the pads stick together in such a way that the entire complex is stable).

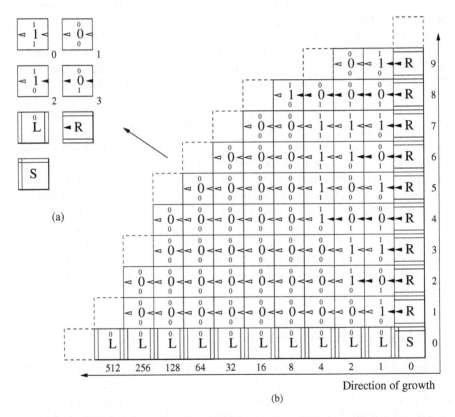

Fig. 3.6. (a) Binary counting tiles. (b) Progression of the growth of the complex

Consider the example depicted in Fig. 3.6. This shows a simple system for counting in binary within the *Tile Assembly Model*. The set of seven different tiles within the system is depicted in Fig. 3.6a: we have four rule tiles (labelled either "1" or "0") r_0, r_1, r_2, r_3, two border tiles (labelled "L" and "R"), and a seed tile ("S"). Sides depicted with a single line have a binding strength of 1; those with a double line have a binding strength of 2. Thick lines depict a binding strength of 0.

We impose the following important restriction: a tile may only be added to the assembly if it is held in place by a *combined* binding strength of 2. In addition, a tile with labelled sides (i.e., a rule tile) may only be added if the labels on its side match those of its proposed neighbor. It is clear that two rule strands in isolation cannot bind together, as the strength of the bond between them can only equal 1. Crystallized by the seed tile, a "scaffold" of L and R tiles emerges to support the assembly, a structure resulting from the binding strengths associated with their sides. Imagine the situation at the beginning of this process, where the assembly consists of one S, one L, and one R tile. The only rule tile that is labelled to match the sides of its neighbors is r_2, so it is added to the complex.

The assembly gradually grows right to left and row by row, with the tiles in row $n > 0$ representing the binary integer n. The growth of the assembly is depicted in Fig. 3.6b, with spaces that *may* be filled by a tile at the next iteration, depicted by the dashed lines. Note that the growth of the complex is limited only by the availability of tiles, and that the "northern" and "western" sides of the assembly are kept exposed as the assembly grows. Notice also how some row n cannot grow left unless row $n - 1$ has grown to at least the same extent (so, for example, a rule tile could not be added to row 2 at the next iteration, because the binding strength would only equal 1).

It has been shown that, for a binding strength of 2, one-dimensional cellular automata can be simulated, and that self-assembly is therefore universal [162]. Work on self-assembly has advanced far beyond the simple example given, and the reader is directed to [163] for a more in-depth description of this. In particular, branched DNA molecules [142] provide a framework for molecular implementation of the model. Double-crossover molecules, with the four sides of the tile represented by "sticky ends", have been shown to self-assemble into a two-dimensional lattice [163]. By altering the binding interactions between different molecules, arbitrary sets of tiles may be constructed.

3.5 Membrane Models

We have already encountered the concept of performing computations by the manipulation of multi-sets of objects. This style of programming is well established in computer science, and Petri nets [129] are perhaps its best known example. Biological systems have provided the inspiration for several multi-set manipulation models, and in this section we describe a few of them.

As we have seen, abstract machines such as the Turing Machine or RAM are widely used in studying the theory of sequential computation (i.e, computation that proceeds by the execution of a single operation per time step). In [30], Berry and Boudol describe a machine that may be used by theorists in the field of *concurrent programming*. This research is concerned with the study of machine models that allow multiple processes executing in parallel. We briefly describe their *Chemical Abstract Machine* (CHAM) in the next section, but first describe its origins.

Most concurrency models are based on architectural principles, for example, networks of processes communicating via "channels" or "threads." In [20], Banâtre and Le Métayer argue that the imposition of architectual control structures in programming languages actually hinders the programmer, rather than helping him or her. They quote Chandy and Misra [41]:

> The basic problem in programming is managing complexity. We cannot address that problem as long as we lump together concerns about the core problem to be solved, the language in which the program is to be written, and the hardware on which the program is to execute. Program development should begin by focusing attention on the problem the be solved and postponing considerations of architecture and language constructs.

Banâtre and Le Métayer argue that it should be possible to construct an abstract, high-level version of a program to solve a given problem, and that that should be free from artificial sequentiality due to architectural constraints. They cite the simple example of finding the maximum element in a nonempty set [21].

In a "traditional", imperative language, the code may look like this. Note that n denotes the size of the set, which is stored in the array $set[]$.

```
maximum ← set[0]
for loop = 1 to n − 1 do begin
    c ← set[loop]
    if c > maximum then maximum ← c
end
```

In this case the program takes the initial maximum to be the first element, and then scans the set in a linear fashion. If an element is found that is larger than the current maximum, then that element *becomes* the current maximum, and so on until the end of the set is reached.

The program imposes a total ordering on the comparisons of the elements when, in fact, the maximum of a set can be computed by performing comparisons in *any* order:

while the set contains $\geq$ two elements **do begin**
    select two elements, compare them and remove the smaller
end

In order to represent such programs, Banâtre and Le Métayer developed the GAMMA language: the General Abstract Model for Multiset Manipulation [20, 21]. In GAMMA, the above statement would be expressed thus:

$$maxset(s) = \Gamma((R, A))(s) \textbf{ where}$$
$$R(x, y) = x \leq y$$
$$A(x, y) = \{y\}$$

The function R specifies the property to be satisfied by selected elements x and y; in this case, x should be less than or equal to y. These elements are then replaced by applying function A. There is no implied order of evaluation, if several disjoint pairs of elements satisfy R, they can be performed in parallel.

Banâtre and Le Métayer recognized that GAMMA programs could almost be viewed as chemical reactions, with the set being the chemical solution, R (the *reaction condition*) a property to be satisfied by reacting elements, and A (the *action*) the resulting product of the reaction. The computation ends when no reactions can occur, and a stable state is reached.

Before we conclude our treatment of GAMMA, let us examine one more program, prime number generation, again taken from [21].

The goal is to produce all prime numbers less than a given N, with $N > 2$. The GAMMA solution removed multiple elements from the multiset $\{2, \ldots, N\}$. We assume the existence of a function $multiple(x, y)$, which returns *true* if x is a multiple of y, and *false* otherwise. The resulting multiset contains all primes less than N:

$$primes(N) = \Gamma((R, A))(\{2 \ldots N\}) \textbf{ where}$$
$$R(x, y) = multiple(x, y)$$
$$A(x, y) = \{y\}$$

One possible "trace" through the computation for $N = 10$ is depicted in Fig. 3.7. The similarities between this approach and the parallel filtering model described earlier are apparent – in the latter model, no ordering is implied when the *remove* operation is performed.

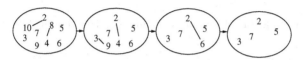

Fig. 3.7. Trace of $primes(10)$

The GAMMA model was extended by Berry and Boudol with the development of the CHAM [30]. The CHAM formalism provides a syntax for molecules as well as a classification scheme for reaction rules. Importantly, it also introduces the *membrane* construct, which is fundamental to the work described in the next section.

P systems

P systems, a variant of the membrane model, were introduced by Gheorge Păun in [118] (see also [119] for an overview of the entire field). They were inspired by features of biological membranes found in nature. These membranes act as barriers and filters, separating the cell into distinct regions and controlling the passage of molecules between regions. However, although P systems were *inspired* by natural membranes, they are not intended to *model* them, and so we refrain here from any detailed discussion of their structure or function.

The membrane structure of a P system is delimited by a *skin* that separates the internals of the system from its outside environment. Within the skin lies a hierarchical arrangement of membranes that define individual *regions*. An *elementary* membrane contains no other membranes, and its region is therefore defined by the space it encloses. The region defined by a nonelementary membrane is the space between the membrane and the membranes contained directly within it. We attach an integer label to each membrane in order to make it addressable during a computation. Since each region is delimited by a unique membrane, we use membrane labels to reference the regions they delimit.

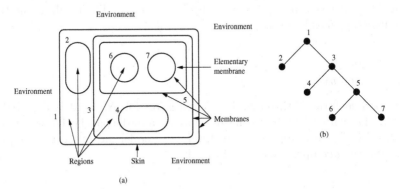

Fig. 3.8. (a) Membrane structure. (b) Tree representation

An example membrane structure is depicted in Fig. 3.8a, with its tree representation in Fig. 3.8b. Note that the skin membrane is represented by the root node, and that leaf nodes represent elementary membranes.

Each region contains a multiset of *objects* and a set of *rules*. Objects are represented by symbols from a given alphabet V. Rules transform or "evolve" objects, and are of the form *before → after*, meaning "evolve every instance of *before* into an instance of *after*." Note that, as we are considering multi-sets, this rule may be applied to multiple objects. Evolution rules are represented by a pair (u, v) of strings over the alphabet V. v may be either v' or $v'\delta$, where δ is a special symbol not in V. v' is a string over $\{a_{here}, a_{out}, a_{in_j}\}$ where j is a membrane identifier. a_{here} means "a copy of a remains in this region", a_{out} means "send a copy of a through the membrane and out of this region", and a_{in_j} means "send a copy of a through the membrane of the region labelled j" (i.e., place a copy of a in membrane j, noting that this is only possible if the current region featuring this rule *contains j*). When the special symbol δ is encountered, the membrane defining the current region (assuming it is not the skin membrane) is "dissolved", and the contents of the current region are placed in the "parent" region (with reference to Fig. 3.8, if membrane 5 were to be dissolved, then region 3 would contain the accessible regions 4, 6 and 7, whereas only regions 4 and 5 are accessible with membrane 5 intact).

In order to simplify the notation, we omit the subscript "here", as it is largely redundant for our purposes. Thus the rule $a → ab$ means "retain a copy of a here and create a copy of b here", whereas $a → b\delta$ means "transform every instance of a into b and then dissolve the membrane." Note that objects on the left hand side of evolution rules are "consumed", or removed, during the process of evaluation.

We may also impose priorities upon rules. This is denoted by $>$, and may be read as follows, using the example $(ff → f) > (f → \delta)$: "transform ff to f as often as possible (halving the number of occurrences of f in the process) until no instances of ff remain, and then transform the one remaining f to δ, dissolving the membrane."

We assume the existence of a "clock" that synchronizes the operation of the system. At each time step, the configuration of the system is transformed by the application of rules in each region in a *nondeterministic, maximally parallel* fashion. This means that objects are assigned to rules nondeterministically in parallel, until no further assignment is possible (hence *maximal*). This series of transitions from one configuration to another forms the computation. A computation halts if no rules may be applied in any region (i.e., nothing can happen). The result of a halting computation is the number of objects sent out through the skin membrane to the outside environment.

We now give a small worked example, taken from [118] and depicted in Fig. 3.9.

This P system calculates n^2 for any given $n \geq 0$. The alphabet $V = \{a, b, d, e, f\}$. Region 3 contains one copy each of objects a and f, and three evolution rules. No objects are present in regions 1 and 2, so no rules can be applied until we reach region 3. We iterate the rules $a → ab$ and $f → ff$ n times in parallel, where $n \geq 0$ is the number we wish to square. This gives n copies of b and 2^n copies of f. We then use $a → b\delta$ instead of $a → ab$, replacing the single a with b and dissolving the membrane. This leaves $n + 1$ copies of b and 2^{n+1} copies of f in region 2; the rules from region 3 are "destroyed" and

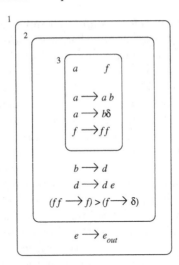

Fig. 3.9. Example P system

the rules from region 2 are now used. The priority relation dictates that we must use the rule $ff \to f$ as often as possible, so in one time step we halve the number of copies of f and, in parallel, transform b^{n+1} to d^{n+1}. In the next step, using $d \to de$, $n+1$ copies of e are produced, and the number of copies of f is once again halved. This step is iterated n times (enforced by the priority relation), producing $n+1$ copies of e at each step, and then $f \to \delta$ dissolves membrane 2. All objects are deposited in the skin membrane, which contains a single rule for e. In one step, all copies of e are sent out of the system using $e \to e_{out}$, the number of copies of e equalling the value of n^2. A trace of the execution of this system for $n = 4$ is given in Fig. 3.10.

Since the initial development of P systems, several variants have appeared in the literature. One of these, described in [116], abandons the use of membrane labels and the ability to send objects to precise membranes, and instead assigns an electrical "charge" to objects and membranes. The motivation behind this change to the basic P system is that sending objects to specifically-labelled membranes is "non-biochemical" and artificial. In the modified P system, objects pass through membranes according to their charge; positively charged objects enter an adjacent negatively charged region (if there are several candidate regions, the transition is selected nondeterministically) and a negatively charged object enters a positively charged region. Neutral objects can only be sent out of the current region; they cannot pass through "inner" membranes. In order to achieve universality, the new variant of the P system is augmented with an additional operation; the action of making a membrane thicker (the membrane dissolving operation is retained). This compensates for the loss of membrane labels by providing an alternative method for controlling the communication of objects through membranes.

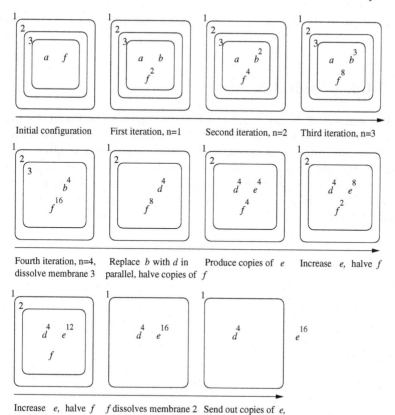

Fig. 3.10. Execution of P system for $n = 4$

In terms of their application, P systems have been used to solve instances of NP-complete problems, notably the Satisfiability and Hamiltonian Path Problems [165]. It is clear that the emerging field of P systems will provide a rich vein of both theoretical results and applications. As Gheorge Păun notes in [117], perhaps the most fundamental open question in the field concerns the physical realization of P systems. Perhaps research in the field of cellular computing (discussed in a later chapter) may provide further insights into this.

3.6 Summary

In this chapter we categorized models of molecular computation into one of four types (filtering, splicing, constructive, and membrane). Abstract models of each type were described, as well as a selection of algorithms within them.

We payed particular attention to the filtering algorithms, especially those of Adleman, Lipton, and Amos et al.

3.7 Bibliographical Notes

The Proceedings of the Second Annual Workshop on DNA Based Computers [94] are significant in that they contain descriptions of both the sticker model and the parallel filtering model, as well as an early paper of Winfree's on self-assembly. Splicing systems are described in detail in [120], a comprehensive review of membrane computing is given in [119], and an assortment of essays dealing mainly with theoretical aspects of DNA computing is collected in [84]. Finally, [13] reviews topics in the theory of DNA computing.

4

Complexity Issues

"Out of intense complexities intense simplicities emerge."
– Sir Winston Churchill.

In this chapter we present an analysis of the complexity and viability of DNA computations. Such analysis has, in part, been motivated by the search for so-called "killer applications": applications of this mode of computation that would establish its superiority within a certain domain. An assured future for DNA computation can only be established through the the discovery of such applications. We introduce our framework for the analysis of DNA algorithms, and argue that existing analyses are flawed and unrealistic. In particular, we argue that computations that are assumed to run in polylogarithmic time actually take polynomial time to realize in the laboratory. We develop further our analysis to motivate the *strong* model of DNA computation, and analyze existing algorithms within it. We argue that our strong model may provide more realistic estimates of the resources required by DNA algorithms. We show how existing models of computation (Boolean circuit and P-RAM) may be effectively simulated using DNA, and give a general framework for the translation of high-level algorithms down to the level of operations on DNA.

4.1 Introduction

Following the initial promise and enthusiastic response to Adleman's seminal work [3] in DNA computation, progress towards the realization of worthwhile computations in the laboratory became stalled. One reason for this is that the computational paradigm employed by Adleman, and generalized by the theoretical work of others [12, 98, 128], relies upon filtering techniques to isolate solutions to a problem from an *exponentially sized* initial solution of DNA. This volume arises because all possible candidate solutions have to be encoded in the initial solution. As Hartmanis points out in [76], the consequence is that, although laboratory computations should work for the smallest problem sizes, the experiments do not realistically scale because vast amounts of DNA are required to initiate computations with even modest problem size. For example, Hartmanis shows that a mass of DNA greater than that of the earth would be required to solve a 200-city instance of the Hamiltonian Path Problem.

The notion of a *"killer application"*, namely, an application of DNA computation that would establish the superiority of this paradigm over others in particular domains, was first explicitly described by Rubin [134], and subsequently by Landweber and Lipton [95]. If practitioners of DNA computation insist on exponential-sized volumes, there can be no hope of discovering "killer applications." An assured future for DNA computation can only be established through the discovery of such applications.

It is not inherently the case that exponentially sized volumes of DNA need be used in DNA computation. Clearly, if exponentially sized volumes are to be avoided, then an alternative algorithmic paradigm to that employed by Adleman in [3] is required. Such a successful paradigm is always likely to emulate traditional computations which *construct* individual solutions rather than sift them out of a vast reservoir of candidates. It might still be argued that the "exponential curse" could not, even then, be avoided for the so-called *NP*-complete problems [63]. If an exact solution is required for any of these, then (employing any extant algorithm) exponential sequential running time is required. A DNA computation, in seeking to reduce this to sub-exponential parallel running time, will certainly require an exponential volume of DNA. However, in general, no one sensibly seeks exact solutions to the *NP*-complete problems. In traditional computation, we either employ heuristics to obtain approximate answers or use randomized methods to obtain exact solutions with high probability. These revised algorithms lead to solutions within polynomial sequential time. Such a view should also be taken for these problems within DNA computation, that is, we should use algorithms which do not inherently require exponential resources.

It is unlikely to be enough, in the quest for "killer applications", to simply have polynomial-volumed computations. We ought, at the same time, to ensure that the vast potential for parallelism is employed to obtain rapid computations. The view taken by the silicon-based parallel computing community [64] is that efficient parallel algorithms, within the so-called Parallel Random Access Machine (P-RAM) model of computation, should have polylogarithmic running time (and use a polynomial number of processors). Problems for which such solutions exist define the complexity class *NC*. If DNA computation is to compete within this domain, then we should clearly also look for polylogarithmic running times within polynomially volumed computations. The discovery of such solutions might well provide candidates for "killer applications." Regardless of the problem considered, it is unlikely to provide a "killer application" unless the computational resources required for a DNA computation (the product of the running time and volume of DNA required) match those needed for a conventional computation (the product of the running time and the number of processors used).

It is clearly crucial, especially when judging the candidacy of a proposed DNA computation for the role of "killer application," to have a firm grasp of the computational resources that it requires. It is often the case that there is not an agreed-upon model of computation in the literature within which we may agree what the required resources are for any particular computation. This chapter attempts to address these issues in a realistic way.

Traditional computational complexity theory [5, 63] is concerned with quantifying the resources (generally *time* and *space*) needed to solve computational problems. Meaningful analysis of the complexity of algorithms may only take place in the context of an agreed-upon model of computation, or *machine model*. Many different machine models have been proposed in the past, including the Deterministic Turing Machine, Boolean circuit [54, 75], and P-RAM [61, 64]. The field of DNA computing also suffers from the problem of proliferation of machine models. Several models have been proposed, within which we may construct algorithms for the solution of computational problems. However, complexity analyses of algorithms within different models of DNA computation are meaningless, since there are no uniform definitions of the concepts of time and space. Furthermore, if we are to compare a DNA-based model with a more traditional machine model, we require a way of demonstrating equivalence between the two.

4.2 An Existing Model of DNA Computation

We may describe abstract models of computation without necessarily considering their implementation. In [64], for example, for the sake of emphasizing what is inherently parallelizable within problems, the authors disregard constraints of implementation. However, in what follows we are naturally constrained by what is feasible in the intended mode of implementation; in this case, what is possible in the molecular biology laboratory. We consider models that operate upon sets of *strings*. It is generally the case that a DNA computation starts and ends with a single set of strings. An algorithm is composed of a sequence of operations upon one or more sets of strings. At the end of the algorithm's execution, a solution to the given problem is encoded as a string in the final set. We use the term *computational substrate* to describe the substance that is acted upon by the implementation of a model. Since DNA is the underlying computational substrate of all models described (as we shall see), we may naturally assume that all abstract models operate on strings over a four-letter alphabet $\{A, G, C, T\}$. For a detailed description of the structure of the DNA molecule, the reader is referred to [156, 155], as well as to Chap. 1.

The weak parallel filtering model

Here we recall the basic legal operations on sets (or *tubes*) [65] within what we now refer to as the *weak* model. The operation set described here is constrained by biological feasibility, but all operations are currently realisable with current technology. The biological implementation of this operation set is described in detail in Chap. 5.

- *remove*$(U, \{S_i\})$. This operation removes from the tube U, in parallel, any string which contains at least one occurrence of any of the substrings S_i.
- *union*$(\{U_i\}, U)$. This operation, in parallel, creates the tube U, which is the set union of the tubes U_i.
- *copy*$(U, \{U_i\})$. In parallel, this operation produces a number of copies, U_i, of the tube U.
- *select*(U). This operation selects an element of U uniformly at random, if U is the empty set then *empty* is returned.

We now describe an example algorithm within the model. The problem solved is that of generating the set of all permutations of the integers 1 to n. The initial set and the filtering out of strings which are not permutations were described earlier. The only non-self-evident notation employed below is $\neg i$ to mean (in this context) any integer in the range which is *not* equal to i.

Problem: Permutations
Generate the set P_n of all permutations of the integers $\{1, 2, \ldots, n\}$.

Solution:

- Input: The input set U consists of all strings of the form $p_1 i_1 p_2 i_2 \ldots p_n i_n$ where, for all j, p_j uniquely encodes "position j" and each i_j is in $\{1, 2, \ldots, n\}$. Thus each string consists of n integers with (possibly) many occurrences of the same integer.

- Algorithm:

    for $j = 1$ to $n - 1$ **do**
        begin
        copy$(U, \{U_1, U_2, \ldots, U_n\})$
        for $i = 1, 2, \ldots, n$ and all $k > j$
                in parallel do remove$(U_i, \{p_j \neg i, p_k i\})$
        union$(\{U_1, U_2, \ldots, U_n\}, U)$
        end
    $P_n \leftarrow U$

- Complexity: $O(n)$ parallel-time.

After the jth iteration of the **for** loop, the computation ensures that in the surviving strings the integer i_j is not duplicated at positions $k > j$ in the string. The integer i_j may be any in the set $\{1, 2, \ldots, n\}$ (which one it is depends in which of the sets U_i the containing string survived). At the end of the computation each of the surviving strings contains exactly one occurrence of each integer in the set $\{1, 2, \ldots, n\}$ and so represents one of the possible permutations. Given the specified input, it is easy to see that P_n will be the set of all permutations of the first n natural numbers. Production of the set P_n can be a useful sub-procedure for other computations, as we have seen in Chap. 3.

Analysis

Attempts have been made to characterize DNA computations using traditional measures of complexity, such as time and space. Such attempts, however, are misleading due to the nature of the laboratory implementation of the computation. We first examine these algorithms from a time complexity standpoint. Most extant models quantify the time complexity of DNA-based algorithms by counting the number of "biological steps" required to solve the given problem. Such steps include the creation of an initial library of strands, separation of subsets of strands, sorting strands on length, and chopping and joining strands.

From the point of view of establishing the parallel-time complexities of algorithms within the model, the basic operations are assumed to take constant time. This assumption has been commonly made by many authors in the literature [65, 98, 113]. However, these operations are frequently implemented in such a way that it is difficult to sustain this claim. For example, the *union* operation consists of pouring a number of tubes into a single tube, and this number is usually, in some way, problem size dependent. Assuming that in general we have a single laboratory assistant, this implies that such operations run in time proportional to the problem size.

Obviously, in the general case, a single laboratory assistant may not pour n tubes into one tube in parallel, nor may he or she split the contents of one tube into n tubes in parallel. This observation, if we are to be realistic in measuring the complexity of DNA computations, requires us to introduce the following constant-time atomic operation:

- *pour*(U, U'). This operation creates a new tube, U, which is the set union of the tubes U and U'.

As we have observed, the *pour* operation is a fundamental component of all compound operations. It therefore follows that more realistic analyses of the time complexities of algorithms may be obtained by taking this operation into consideration.

4.3 A Strong Model of DNA Computation

In what follows we refine the weak model just described. We assume that the initial tube (which takes at most linear time to construct) is already set up.

The *pour* operation is fundamental to all compound operations within our weak model. We must therefore reassess the time complexity of these operations. The *remove* operation requires adding to U

1. i tubes containing short *primers* to mark selected strands
2. restriction enzymes to destroy marked strands

This operation is inherently sequential, since there must be a pause between steps 1 and 2 in order to allow the primers to mark selected strands. Therefore, the *remove* operation takes $O(i)$ time. Creating the *union* of i tubes is an inherently sequential operation, since the assistant must first pour U_1 into U, then U_2, and so on, up to U_i. Rather than taking constant time, the *union* operation actually takes $O(i)$ time. It is clear that the *copy* operation may be thought of as a *reverse-union* operation, since the contents of a single tube U are split into many tubes, $\{U_i\}$. Therefore, *copy* takes $O(i)$ time.

We may strengthen the strong model even further by taking kinetic issues into consideration. All filtering models rely upon the fact that every possible solution to the given problem is present in the initial tube with equal probability. These solutions are created by the action of component molecules "sticking" together to form long chains. Given a particular problem of size n, we may naively assume that the creation of the initial library takes $O(\log n)$ time [93]. Given a particular solution, we may imagine a balanced binary tree, where each leaf node represents a component molecule, an internal node represents a "sticking-together" reaction, and the root node represents the string encoding the solution. Since the depth of the tree is $\log n$, the time complexity of the creation phase immediately follows.

However, Kurtz et al.[93] present a more detailed analysis of the kinetic issues involved, suggesting that the creation phase actually takes $\Omega(n^2)$ time. Given the space available, and the fact that in what follows we assume that the initial library is set up, we omit a detailed description of the arguments presented in [93]. However, we acknowledge the importance of taking kinetic issues into consideration in the future.

Complexity comparisons

In this section we compare time complexities for algorithms previously described [65] within both the weak and strong models. In particular, we examine in detail the problem of generating a set of permutations. This will characterize the general form of comparisons that can be made, so that in the space available we merely tabulate comparisons for other algorithms.

In [65] the authors claimed a time complexity for the **Permutations** algorithm of $O(n)$. This is based on the assumption that *remove* is a constant-time operation. We justify a new time complexity of $O(n^2)$ as follows: at each iteration of the **for** loop we perform one *copy* operation, n *remove* operations and one *union* operation. The *remove* operation is itself a compound operation, consisting of $2n$ *pour* operations. The *copy* and *union* operations consist of n *pour* operations.

Similar considerations cause us to reassess the complexities of the algorithms described in [65], according to Table 4.1.

Table 4.1. Time comparison of algorithms within the weak and strong models

Algorithm	Weak	Strong
Three coloring	$O(n)$	$O(n^2)$
Hamiltonian path	$O(1)$	$O(n)$
Subgraph isomorphism	$O(n)$	$O(n^2)$
Maximum clique	$O(n)$	$O(n^2)$
Maximum independent set	$O(n)$	$O(n^2)$

Although we have concentrated here on adjusting time complexities of algorithms described in [65], similar adjustments can be made to other work. An example is given in the following section.

4.4 Ogihara and Ray's Boolean Circuit Model

Several authors [50, 113, 128] have described models of DNA computation that are Turing-complete. In other words, they have shown that any process that could naturally be described as an algorithm can be realized by a DNA computation. [50] and [128] show how any Turing Machine computation may be simulated by the addition of a *splice* operation to the models already described in this book. In [113], Ogihara and Ray describe the simulation of Boolean circuits within a model of DNA computation. The complexity of these simulations is therefore of general interest. We first describe that of Ogihara and Ray [113].

Boolean circuits are an important Turing-equivalent model of parallel computation (see [54, 75]). An n-input *bounded fan-in* Boolean circuit may be viewed as a directed, acyclic graph, S, with two types of node: n *input* nodes with in-degree (i.e., input lines) zero, and *gate* nodes with maximum in-degree two. Each input node is associated with a unique Boolean variable x_i from the input set $X_n = (x_1, x_2, \ldots, x_n)$. Each gate node, g_i is associated with some Boolean function $f_i \in \Omega$. We refer to Ω as the circuit *basis*. A *complete* basis is a set of functions that are able to express all possible Boolean functions. It is well known [143] that the NAND function provides a complete basis by itself, but for the moment we consider the common basis, according to which

$\Omega = \{\wedge, \vee, \neg\}$. In addition, S has some unique *output* node, s, with out-degree zero. An example Boolean circuit for the three-input *majority function* is depicted in Fig. 4.1.

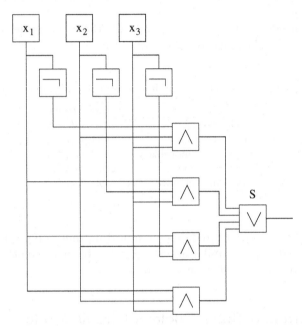

Fig. 4.1. Boolean circuit for the three-input majority function

The two standard complexity measures for Boolean circuits are *size* and *depth*: the size of a circuit is S, and m is the number of gates in S; its depth, d, is the number of gates in the *longest* directed path connecting an input vertex to an output gate. The circuit depicted in Fig. 4.1 has size 8 and depth 3.

In [113], Ogihara and Ray describe the simulation of Boolean circuits within a model of DNA computation. The basic structure operated upon is a *tube*, U, which contains strings representing the results of the output of each gate at a particular depth. The initial tube contains strands encoding the values of each of the inputs X_n.

In what follows, $\Omega = \{\wedge, \vee\}$. For each i, $1 \leq i \leq m$, a string $\sigma[i]$ is fixed. The presence of $\sigma[i]$ in U signifies that g_i evaluates to 1. The absence of $\sigma[i]$ in U signifies that g_i evaluates to 0. The initial tube, U, (i.e., a tube representing the inputs X_n is created as follows:

> **for each** gate x_i **do**
>> **if** $x_i = 1$ **then** $U \leftarrow U \cup \sigma[i]$
> **end for**

The simulation of gates at level $k > 0$ proceeds as follows. We denote by i_1 and i_2 the indices of the gates that supply the inputs for g_i.

(1) Input(U)
(2) **for** $k = 1$ to d **do**
(3)   **for** each gate g_i at level k **in parallel do**
(4)   **if** g_i computes $\vee$ **then do**
(5)   **if** $(\sigma[i_1] \in U)$ **or** $(\sigma[i_2] \in U)$ **then** $U \leftarrow U \cup \sigma[i]$
(6)   **else if** g_i computes $\wedge$ **then do**
(7)   **if** $(\sigma[i_1] \in U)$ **and** $(\sigma[i_2] \in U)$ **then** $U \leftarrow U \cup \sigma[i]$
(8)   **end for**
(9) **end for**

At step (1) the initial tube, U, is created. Then, for each circuit level $k > 0$ (step 2), the gates at level k are simulated in parallel (steps 3 through 8). The simulation of each gate g_i at level k is achieved as follows. If g_i is an $\vee$-gate, the string $\sigma[i]$ is made present[1] in U (i.e., g_i evaluates to 1) if either of the strings representing the inputs to g_i is present in U (step 5). If g_i is an $\wedge$-gate, the string $\sigma[i]$ is made present in U (i.e., g_i evaluates to 1) if *both* of the strings representing the inputs to g_i are present in U (step 7). The simulation then proceeds to the next level. At the termination of the computation, Ogihara and Ray analyze the contents of U to determine the output of the circuit.

4.4.1 Ogihara and Ray's Implementation

In this section we describe the laboratory implementation of the abstract Boolean circuit model just described. The circuit simulated in [113] is depicted in Fig. 4.2.

Each gate g_i is assigned a sequence of DNA, $\sigma[i]$, of length $\mathcal{L}$, beginning with a specific restriction site, $\mathcal{E}$. Each edge $i \to j$ is assigned a sequence $e_{i,j}$ that is the concatenation of the complement of the 3' $\mathcal{L}/2$-mer of $\sigma[i]$ and the complement of the 5 $\mathcal{L}/2$-mer of $\sigma[j]$. In this way, $e_{i,j}$ acts as a "splint" between g_i and g_j if and only if both $\sigma[i]$ and $\sigma[j]$ are present. However, we later highlight a case where this strand design does not hold.

The simulation of gates at level 0 (i.e., the construction of the initial tube) proceeds as follows. Begin with a tube of solution containing no DNA. For each input gate x_i that evaluates to 1, pour into the tube a population of strands representing $\sigma[i]$. If x_i evaluates to 0, do not add $\sigma[i]$ to the tube.

We now consider the simulation of gates at level $k > 0$. The simulation of $\vee$-gates differs from that $\wedge$-gates, so we first consider the case of an $\vee$-gate, g_j, at level k.

First, pour into the tube strands representing $\sigma[j]$. Then, for each edge $i \to j$ pour into the tube strands representing $e_{i,j}$. Allow ligation to occur. If strands

---

[1] The process by which this is achieved is described in detail in Sect. 4.4.1.

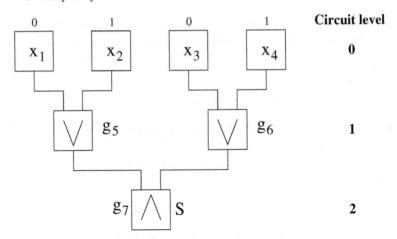

Fig. 4.2. Boolean circuit simulated by Ogihara and Ray

representing either of the inputs to g_j are present, the edge strands will cause strands of length $2\mathcal{L}$ to be formed, consisting of the concatenation of the sequence representing the g_i with the sequence representing g_j. This process is depicted in Fig. 4.3a. These strands are then separated out by running the solution on a polyacrylamide gel. These strands are then cut with a restriction enzyme recognizing sequence $\mathcal{E}$. This step leaves in solution strands of length $\mathcal{L}$ that correspond to g_j (i.e., g_j evaluates to 1).

We now consider the simulation of an $\wedge$-gate, g_j, at level k. Again, pour into the tube strands representing $\sigma[j]$. Then, for all edges $i \to j$ pour into the tube strands representing $e_{i,j}$. Allow ligation to occur. If strands representing *both* of the inputs to g_j are present, the edge strands will cause strands of length $3\mathcal{L}$ to be formed, consisting of the concatenation of the sequence representing the first input to g_j, the sequence representing g_j, and the sequence representing the second input to g_j (see Fig. 4.3b). Note that this splinting only occurs if the polarity of the edge splints is designed carefully, and we consider this in the next section. The strands of length $3\mathcal{L}$ are again separated out by a gel and cut with the appropriate restriction enzyme. This step leaves in solution strands of length $\mathcal{L}$ that correspond to g_j (i.e., g_j evaluates to 1).

If $k = d$, we omit the gel electrophoresis and restriction stages, and simply run the solution on a gel. If the output gate, s, is an $\vee$-gate, the output of the circuit is 1 if and only if $2\mathcal{L}$ length strands exist in solution. If s is an $\wedge$-gate, the output of the circuit is 1 if and only if $3\mathcal{L}$ length strands exist in solution.

Experimental results obtained

In [113], Ogihara and Ray report attempts to simulate the circuit previously described using techniques of molecular biology. Their implementation is as

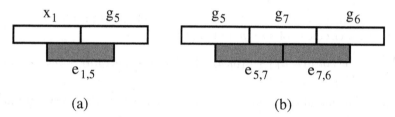

Fig. 4.3. (a) Splinting for ∨-gate. (b) Splinting for ∧-gate

described in the previous section. The results obtained were ambiguous, although Ogihara and Ray claim to have identified the ambiguity as being caused by pipetting error.

Complexity analysis

Within the strong model, it appears that the methods of [113] require a number of *pour* operations where linearity in the size of the circuit simulated is easily demonstrated. We recall that the basic circuit model considered in their paper is that of "semi-unbounded fan-in circuits":

Definition 1. *A semi-unbounded fan-in Boolean circuit of n inputs is a labelled, directed acyclic graph whose nodes are either inputs or gates. Inputs, of which there are exactly 2n, have fan-in 0 and each is labelled with a unique Boolean literal x_i or $\overline{x_i}$ (1 ≤ i ≤ n). Gates are either ∧ (conjunction) or ∨ (disjunction) gates. The former have fan-in of exactly 2; the latter may have an arbitrary fan-in. There is a unique gate with fan-out of 0, termed the output. The depth of a gate is the length of the longest directed path to it from an input. The size of such a circuit is the number of gates; its depth is the depth of the output gate.*

We note, in passing, that there is an implicit assumption in the simulation of [113] that such circuits are *levelled*, i.e. every gate at depth k $(k > 0)$ receives its inputs from nodes at depth k-1. While this is not explicitly stated as being a feature of their circuit model, it is well known that circuits not organized in this way can be replaced by levelled circuits with at most a constant factor increase in size.

   The simulation associates a unique DNA pattern, $\sigma[i]$, with each node of the circuit, the presence of this pattern in the pool indicating that the ith node evaluates to 1. For a circuit of depth d the simulation proceeds over d rounds: during the kth round only strands associated with nodes at depth k-1 (which evaluate to 1) and gates at depth k are present. The following *pour* operations are performed at each round: in performing round k $(k \geq 1)$ there is an operation to pour the strand $\sigma[g_i]$ for each gate g_i at depth k. Furthermore, there are operations to pour a "linker" for each different *edge* connecting a node at level $k-1$ to a node at level k. We thus have a total, in simulating a circuit C of size m and depth d, of

$$\sum_{k=1}^{d} | \{g : depth(g) = k\} | + | \{(g,h) : depth(h) = k \; and \; (g,h) \in Edges(C)\} |$$

distinct pour operations being performed. Obviously (since every gate has a unique depth)

$$\sum_{k=1}^{d} | \{g : depth(g) = k\} | = m$$

Furthermore,

$$\sum_{k=1}^{d} | \{g : depth(g) = k\} | + | \{(g,h) : depth(h) = k \; and \; (g,h) \in Edges(C)\} | \geq 2m$$

since every gate has *at least* two inputs. It follows that the total number of *pour* operations performed over the course of the simulation is at least $3m$.

Despite these observations, Ogihara and Ray's work is important because it establishes the Turing-completeness of DNA computation. This follows from the work of Fischer and Pippenger [60] and Schnorr [139], who described simulations of Turing Machines by combinational networks. Although a Turing Machine simulation using DNA has previously been described by Reif [128], Ogihara and Ray's method is simpler, if less direct.

4.5 An Alternative Boolean Circuit Simulation

Since it is well known [54, 75, 157] that the NAND function provides a complete basis by itself, we restrict our model to the simulation of such gates. In fact, the realization in DNA of this basis provides a far less complicated simulation than using other complete bases. It is interesting to observe that the fact that NAND offers the most suitable basis for Boolean network simulation within DNA computation continues the traditional use of this basis as a fundamental component within new technologies, from the work of Sheffer [143], that established the completeness of NAND with respect to propositional logic, through classical gate-level design techniques [75], and, continuing, in the present day, with VLSI technologies both in nMOS [106], and CMOS [159, pp. 9–10].

The simulation proceeds as follows: An n-input, m-output Boolean network is modelled as a directed acyclic graph, $S(V, E)$, in which the set of vertices V is formed from two disjoint sets, X_n, the *inputs* of the network (of which there are exactly n) and G, the *gates* (of which exactly m are distinguished as output gates). Each input vertex has in-degree 0 and is associated with a single Boolean variable, x_i. Each gate has in-degree equal to 2 and is associated with the Boolean operation NAND. The m distinguished output gates, $t_1, t_2, \ldots, t_m$, are conventionally regarded as having out-degree equal to 0. An assignment of Boolean variables from $< 0,1 >^n$ to the inputs X_n ultimately induces Boolean values at the output gates $< t_1, \ldots, t_m >$. An n-input, m-output Boolean network S is said to compute an n-input, m-output Boolean function

$$f(X_n) :< 0,1 >^n \rightarrow < 0,1 >^m =_{def} < f^{(i)}(X_n) :< 0,1 >^n \rightarrow \{0,1\} : 1 \leq i \leq m > \text{ if } \forall \alpha \in < 0,1 >^n \; \forall 1 \leq i \leq m \; t_i(\alpha) = f^{(i)}(\alpha).$$

The two standard complexity measures for Boolean networks are *size* and *depth*: the size of a network S, denoted $C(S)$, is the number of gates in S; its depth, denoted by $D(S)$, is the number of gates in the *longest* directed path connecting an input vertex to an output gate.

The simulation takes place in three distinct phases:

1. Set-up
2. Level simulation
3. Final read-out of output gates

We now describe each phase in detail.

Set-up

In what follows we use the term *tube* to denote a set of strings over some alphabet σ. We denote the jth gate at level k by g_k^j. We first create a tube, T_0, containing unique strings of length l, each of which corresponds only to those input gates that have value 1. We then create, for each level $1 \leq k < D(S)$, a tube T_k containing unique strings of length $3l$ representing each gate at level k. We also create a tube S_k, containing strings corresponding to the complement of positions $2l - 5$ to $2l + 5$ for each g_k^j. We define the concept of complementarity in the next section, but for the moment we assume that if sequence x and its complement $\bar{x}$ are present in the same tube, the string containing sequence x is in some way "marked."

We then create tube $T_{D(S)}$, containing unique strings representing the output gates $< t_1, \ldots, t_m >$. These strings representing gates at level $1 \leq k < D(S)$ are of the form $x_k^j y_k^j z_k^j$. If gate g_k^j takes its input from gates g_{k-1}^m and g_{k-1}^n, then the sequence representing x_k^j is the complement of the sequence representing z_{k-1}^m, and y_j^k is the complement of the sequence representing z_{k-1}^n. The presence of z_k^j therefore signifies that g_k^j has an output value of 1.

The strings in tube $T_{D(S)}$ are similar, but the length of the sequence $z_{D(S)}^j$ is in some way proportional to j. Thus, the length of each string in $T_{D(S)}$ is linked to the index of the output gate it represents.

Level simulation

We now describe how levels $1 \leq k < D(S)$ are simulated. We create the set
union of tubes T_{k-1} and T_k. Strings representing gates which take either of
their inputs from a gate with an output value of 1 are "marked", due to their
complementary nature. We then remove from T_k all strings that have been
marked twice (i.e., those representing gates with both inputs equal to one).
We then split the remaining strings after section y_k^j, retaining the sequences
representing z_k^j. This subset then forms the input to tube T_{k+1}.

Final read-out of output gates

At level D(S) we create the set union of tubes $T_{D(S)-1}$ and $T_{D(S)}$ as described
above. Then, as before, we remove from this set all strings that have been
marked twice. By checking the length of each string in this set we are therefore
able to say which output gate has the value 1 and which has the value zero,
respectively, by the presence or absence of a string representing the gate in
question.

4.6 Proposed Physical Implementation

We now describe how the abstract model detailed in the previous section
may be implemented in the laboratory using standard biomolecular manipulation techniques. The implementation is similar to that of our *parallel filtering
model*, described in [6, 14].

   We first describe the design of strands representing the input gates X_n.
For each X_n that has the value 1 we synthesize a unique strand of length l.
We now describe the design of strands representing gates at level 1. We have
already synthesized a unique strand to represent each g_k^j at the set-up stage.
Each strand is comprised of three components of length l, representing the
gate's inputs and output. Positions 0 to l represent the first input, positions
$l + 1$ to $2l$ represent the second input, and positions $2l + 1$ to $3l$ represent the
gate's output. Positions $l - 3$ to $l + 3$ and positions $2l - 3$ to $2l + 3$ correspond
to the restriction site $CACGTG$. This site is recognized and cleaved exactly
at its midpoint by the restriction enzyme $PmlI$, leaving blunt ends. Due to
the inclusion of these restriction sites, positions 0 to 2, $l+1$ to $l+3$, and $2l+1$
to $2l + 3$ correspond to the sequence GTG, and positions $l - 3$ to l, $2l - 3$
to $2l$, and $3l - 3$ to $3l$ correspond to the sequence CAC. The design of the
other subsequences is described in Section 4.5. A graphical depiction of the
structure of each gate strand is shown in Fig. 4.4.

   The simulation proceeds as follows for levels $1 \leq k < D(S)$.

1. At k pour the strands in tube T_{k-1} into T_k. These anneal to the gate
   strands at the appropriate position.

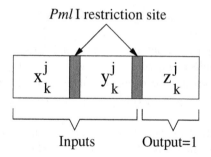

Fig. 4.4. Structure of a gate strand

2. Add ligase to T_k in order to seal any "nicks."
3. Add the restriction enzyme $PmlI$ to T_k. Because of the strand design, the enzyme cleaves only those strands that have *both* input strands annealed to them. This is due to the fact that the *first* restriction site $CACGTG$ is only made fully double stranded if both of these strands have annealed correctly. This process is depicted in Fig. 4.5.
4. Denature the strands and run T_k through a gel, retaining only strands of length $3l$. This may be achieved in a single step by using a denaturing gel [136].
5. Add tube S_k to tube T_k. The strands in tube S_k anneal to the *second* restriction site embedded within each retained gate strand.
6. Add enzyme $PmlI$ to tube T_k, which "snips" off the z_k^j section (i.e., the output section) of each strand representing a retained gate.
7. Denature and run T_k through another gel, this time retaining only strands of length l. This tube, T_k, of retained strands, forms the input to T_{k+1}. We now proceed to the simulation of level $k + 1$.

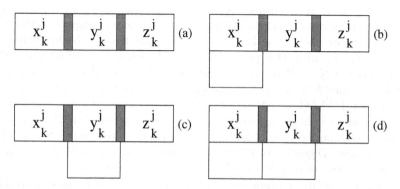

Fig. 4.5. (a) Both inputs = 0. (b) First input = 1. (c) Second input = 1. (d) Both inputs = 1

At level $D(S)$ we carry out steps 1 through 7, as described above. However, at steps 4 and 7 we retain all strands of length $\geq 3l$. We are now then ready to implement the final read-out phase. This involves a simple interpretation of the final gel visualisation. Since we know the unique length u_j of each $z_{D(S)}^j$ section of the strand for each output gate t_j, the presence or absence of a strand of length $u_j + 2l$ in the gel signifies that t_j has the value 1 or 0 respectively.

We now illustrate this simulation with a small worked example. We describe the DNA simulation of the small NAND-gate circuit depicted in Fig. 4.6a. The sequences chosen to represent each gate strand are represented in Fig. 4.6b, the shape of each component diagrammatically representing its sequence. It is clear from the diagram that the input and output components of connected gates fit each other in a "lock-key" fashion. Given the inputs shown in Fig. 4.6a, the simulation proceeds as follows. Gate g_1 has input values of 0 and 1, so we add to tube T_0 a strand representing the complement of the second input component of the strand representing g_1. Gate g_2 has both inputs equal to 1, so we add to tube T_0 strands representing the complement of both input components of g_2. We then add tube T_0 to tube T_1, allow annealing, and ligate. We next add the appropriate restriction enzyme. The strands representing g_1 are not cleaved, since the first restriction site is not made fully double stranded. The strands representing g_2 *are* cleaved, since the first restriction site is fully double stranded. We then sort T_1 on length using a gel following PCR, "snip" off the output components of full-length strands using a second blunt ended restriction enzyme, and remove the output components. These retained strands are then added as input to T_2, and we repeat the above procedure. After sorting on length, the presence of intact, full-length strands indicates that the circuit has the output value 1 (Fig. 4.6c).

However, there is a potential problem in the restriction of side-by-side annealed oligonucleotides in that there is a nick in the double stranded DNA. This problem can be potentially solved by using DNA ligase. The problem is that it is difficult to ensure 100% ligation. Failure of ligation yields false positives, i.e., strands that should have been restricted escape into the next stage. Therefore the restriction enzyme used will have to be indifferent to the presence of a nick. One potential group of such nick resistant restriction enzymes is the type III restriction enzymes. The type III enzymes cut at a specific distance 3' from the recognition site; e.g., restriction enzyme Eco57I recognizes CTGAAG, and then cuts the top 16 nucleotides and the bottom 14 nucleotides of the recognition site. We may engineer Eco57I sites in all the input oligonucleotides such that the restriction site extends across adjacent sequences. Digestion of the DNA should be possible only if both input oligonucleotides are double stranded. We can easily test nick resistance and dependence on double strandedness of both input oligonucleotides by creation of artificial substrates with and without nicks and substrates that are fully double stranded, half double stranded, and fully single stranded.

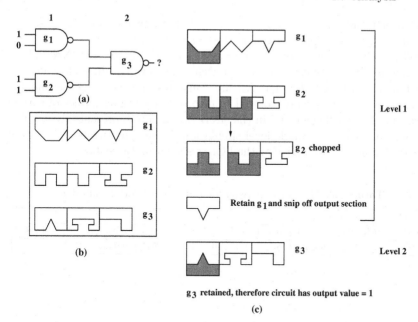

Fig. 4.6. (a) Circuit to be simulated. (b) Strand representing gates. (c) Stages of the simulation

4.7 Analysis

We now compare both our and Ogihara and Ray's models by describing how Batcher sorting networks [22] may be implemented within them. Batcher networks sort n inputs in $O(log^2 n)$ stages. In [157], Wegener showed that if $n = 2^k$ then the number of *comparison* modules is $0.5n(\log n)(\log n - 1) + 2n - 2$. The circuit *depth* (again expressed in terms of the number of comparison modules) is $0.5(\log n)(\log n + 1)$. A comparison module has two (Boolean) inputs, x and y, and two outputs, $MIN(x, y)$ (which is just x AND y) and $MAX(x, y)$ (which is just x OR y).

Using NAND we can build a comparison module with five NAND gates and having depth 2 (the module is levelled; so since the Batcher network is levelled with respect to comparison modules, the whole realization in NAND gates will be levelled). The NAND gate realization is

$MIN(x, y) = NAND(NAND(x, y), NAND(x, y))$ - 2 gates, depth 2;
$MAX(x, y) = NAND(NAND(x, x), NAND(y, y))$ - 3 gates, depth 2;

If we assume that $n = 2^k$ we get the total size (in terms of number of gates) as $2.5(\log n)(\log n - 1) + 10n - 10$ and depth (in gates) as $(\log n)(log n + 1)$.

Ogihara and Ray use an AND and OR gate simulation, so they would realize a comparison module with two gates ($MAX(x, y) = x$ OR y; $MIN(x, y) =$

x AND y)) and depth 1, giving the size of the network as $n(\log n)(\log n - 1) + 4n - 4$ and the depth of the network as $0.5(\log n)(\log n + 1)$.

Within the context of our *strong model* [10], the volumes of DNA and the time required to simulate an n-input Batcher network within each model are depicted in Table 4.2. $K1$ and $K2$ are constants, representing the number of copies of a single strand required to give reasonable guarantees of correct operation. The coefficient of 7 in the A&D time figure represents the number of separate stages in a single level simulation. If we concentrate on the time measure for $n = 2^k$ we arrive at the figures shown in Table 4.3.

Table 4.2. Model comparison for Batcher network simulation

Model	Volume	Time
O&R	$(K1)(n(\log n)(\log n - 1) + 4n - 4)$	$n(\log n)(\log n - 1) + 4n + 4$
A&D	$(K2)(2.5(\log n)(logn - 1) + 10n - 10)$	$7(\log n)(\log n + 1)$

Table 4.3. Time comparisons for different values of k

n	k	$O\&R$	$A\&D$
1024	10	92196	770
2^{20}	20	$4 * 10^8$	2940
2^{40}	40	$1.7 * 10^{15}$	11480

Roweis et al. claim that their *sticker model* [133] is feasible using 2^{56} distinct strands. We therefore conclude that our implementation is technically feasible for input sizes that could not be physically realized *in silico* using existing fabrication techniques.

4.8 Example Application: Transitive Closure

We now demonstrate how a particular computation, *transitive closure*, may be translated into DNA via Boolean circuits. In this way we demonstrate the feasibility of converting a general algorithm into a sequence of molecular steps.

The transitive closure problem

The computation of the *transitive closure* of a directed graph is fundamental to the solution of several other problems, including shortest path and connected components problems. Several variants of the transitive closure problem exist;

we concentrate on the *all-pairs transitive closure problem*. Here, we find all pairs of vertices in the input graph that are connected by a path.

The transitive closure of a directed graph $G = (V, E)$ is the graph $G^* = (V, E^*)$, where E^* consists of all pairs $< i, j >$, such that either $i = j$ or there exists a path from i to j. An example graph G is depicted in Fig. 4.7a with its transitive closure G^* depicted in Fig. 4.7b.

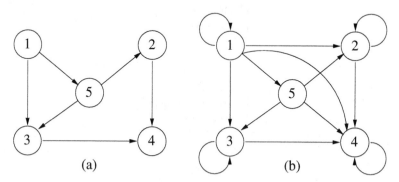

Fig. 4.7. (a) Example graph, G. (b) Transitive closure of G

We represent G by its adjacency matrix, A. Let A^* be the adjacency matrix of G^*. In [83], JàJà shows how the computation of A^* may be reduced to computing a power of a Boolean matrix.

We now describe the structure of a $NAND$ gate Boolean circuit to compute the transitive closure of the $n \times n$ Boolean matrix A. For ease of exposition we assume that $n = 2^p$.

The transitive closure, A^*, of A is equal to $(A + I)^n$, where I is the $n \times n$ identity matrix. This is computed by $p = \log_2 n$ levels. The ith level takes as input the matrix output by level $i - 1$ (with level 1 accepting the input matrix $A + I$) and squares it: thus level i outputs a matrix equal to $(A + I)^{2^i}$ (Fig. 4.8).

To compute A^2 given A, the n^2 Boolean values $A^2_{i,j}$ $(1 \leq i, j \leq n)$ are needed. These are given by the expression $A^2_{i,j} = \bigvee_{k=1}^{n} A_{i,k} \wedge A_{k,j}$. First, all the n^3 terms $A_{i,k} \wedge A_{k,j}$ (for each $1 \leq i, j, k \leq n$) are computed in two (parallel) steps using two $NAND$ gates for each $\wedge$-gate simulation. Using $NAND$ gates, we have $x \wedge y = NAND(NAND(x, y), NAND(x, y))$. The final stage is to compute all of the n^2 n-bit sums $\bigvee_{k=1}^{n} (A_{i,k} \wedge A_{k,j})$ that form the input to the next level.

The n-bit sum $\bigvee_{k=1}^{n} x_k$, can be computed by a $NAND$ circuit comprising p levels each of depth 2. Let level 0 be the inputs $x_1, \ldots, x_n$; level i has 2^{p-i} outputs $y_1, \ldots, y_r$, and level $i + 1$ computes $y_1 \vee y_2, y_3 \vee y_4, \ldots, y_{r-1} \vee y_r$.

Input matrix:$A+I$

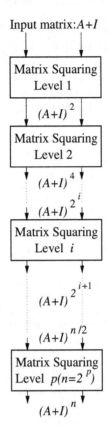

Fig. 4.8. Matrix squaring

Each $\vee$ can be computed in two steps using three $NAND$ gates, since $x \vee y = NAND(NAND(x,x), NAND(y,y))$.

In total we use $2p^2$ parallel steps and a network of size $5p2^{3p} - p2^{2p}$ $NAND$ gates, i.e., of $2(\log_2 n)^2$ depth and of $5n^3 \log_2 n - 2n^2 \log_2 n$ size.

We now consider the biological resources required to execute such a simulation. Each gate is encoded as a single strand; in addition, at level $1 \leq k \leq D(S)$ we require an additional strand per gate for removal of output sections. The total number of distinct strands required is therefore $n(-1 + 2 \log n^2(-1 + 5kn))$, where k is a constant, representing the number of copies of a single strand required to give a reasonable guarantee of correct operation.

4.9 P-RAM Simulation

We now describe, after [11], our DNA-based P-RAM simulation. At the coarsest level of description, the method simply describes how any so-called P-RAM algorithm may be emulated in DNA. The P-RAM is a widely used model in the design of efficient parallel algorithms, especially by the international community of theorists working in the field. There are literally thousands of published algorithms within this model which, often employing ingenious combinatorics, are remarkably efficient. One of the attractions of our approach is that we can call upon this vast published expertise. The immediate difficulty however is that it is not clear how to *directly* emulate the P-RAM in DNA without invoking great intricacy and unrealistic bench chemistry. On the other hand, we have been able to describe how a Boolean circuit may be efficiently emulated in DNA. Although both models are universal (that is, are Turing-complete), it is in general very difficult to "program" a Boolean circuit (that is, to design such a circuit to solve a specific problem) without a great degree of low level intricacy and opacity. The reverse is true for the P-RAM which yields a natural, high-level, programming environment. The benefit of our methodology is that we are able to describe a general method for compiling P-RAM algorithms into Boolean Combinational Logic Networks.

Overall, then, we believe that we have a theoretical model which is easy to program, may call upon a vast literature of efficient parallel algorithmic design, and which (through Boolean circuit emulation) has the prospect of emulation in DNA using relatively simple and clean processes. Moreover, and very importantly, there is only a logarithmic loss in efficiency in the emulation process. We now outline, in greater technical detail, how the emulation works.

The work presented in [8] gives a simulation of Boolean networks by DNA algorithms that requires parallel time comparable to the network depth and volume comparable to the network size. This follows the work of Ogihara and Ray [113], in which they describe a Boolean circuit model that, within our *strong model* of DNA computation [10], runs in parallel time proportionate to the circuit *size*. Although we point out in [8] that a simple modification to Ogihara and Ray's simulation would achieve the same time complexity of our simulation, we believe that the biological implementation of our model is more straight forward. We justify this remark later.

The crucial extension of our simulation is presented in [9]: this presents a full detailed translation from the high-level P-RAM algorithms into DNA, the translation being accomplished by using a Boolean network representation of the P-RAM algorithm as an intermediate stage. In total this translation achieves the following performance: if A is a P-RAM algorithm using $P(n)$ processors, taking $S(n)$ memory space, and taking $T(n)$ time, then the total computation time of the DNA algorithm is bounded by $O(T(n)logS(n))$, and the total volume of DNA used is $O(P(n)T(n)S(n)logS(n))$. As a consequence this gives a direct translation from any NC algorithm into an effective DNA algorithm. The simulation presented thereby moves the purely theoretical "in principle" result that DNA can realize NC "efficiently" into the realm where such algorithms can be realized in DNA by a practical, universal translation process. In [9] a formal definition of the low-level instruction set available to each processor in the P-RAM is specified: this provides standard memory access facilities (load and store), basic arithmetic operations, and conditional branching instructions. The compilation into DNA involves three basic stages: compiling the P-RAM program down to a low-level sequence of instructions; translating this program into a combinational logic network; and, finally, encoding the actions of the resulting Boolean networks into DNA. The actions performed in the second stage lie at the core of the translation. For each processor invoked at the kth parallel step, there will be *identical* (except for memory references) sequences of low-level instructions; the combinational logic block that simulates this (parallel) instruction takes as its input all of the bits corresponding to the current state of the memory, and produces as output the same number of bits representing the memory contents after execution has completed. Effecting the necessary changes involves no more than employing the required combinational logic network to simulate the action of any low-level instruction. The main overhead in the simulation comes from accessing a specific memory location, leading to the $log S(n)$ slow-down. All other operations can be simulated by well established combinational logic methods without any loss in runtime. It is well known that the NAND function provides a complete basis for computation by itself; therefore we restrict our model to the simulation of such gates. In fact, the realization in DNA of this basis provides a far less complicated simulation than using other complete bases.

We now describe the simulation in detail. This detailed work is due mainly to the second author of [9], and is used with permission. We provide a full, formal definition of the CREW P-RAM model that we will use. This description will consider processors, memory, the control regime, memory access rules, and the processor instruction set. Finally we describe the complexity measures that are of interest within this model.

The p processors, labelled $P_1, P_2, \ldots, P_p$ are identical, and can execute instructions from the set described below. Processors have a unique identifier: that of P_i being i. The global common memory, M, consists of t locations $M_1, M_2, \ldots, M_t$, each of which is exactly r-bits long. The initial input data consists of n items, which are assumed to be stored in locations $M[1], \ldots, M[n]$ of the global memory. The main control program, C, sequentially executes instructions of the form

$$k; \textbf{for } x \in L_k \textbf{ par do } inst_k(x)$$

L_k is a set of processor identifiers. For each processor P_x in this list the instruction specified by $inst_x$ is executed. Each processor executes the same sequence of instructions, but on different memory locations. The control program language augments the basic parallel operation command with the additional control flow statements

$$\textbf{for } counter\text{-}name \leftarrow start\text{-}value \textbf{ step } step\text{-}value \textbf{ to } end\text{-}value \textbf{ do}$$
$$instruction\text{-}sequence$$

and

> **repeat** *repeat-value* **times**
> *instruction-sequence*

There is a restriction that the variables *start-value, step-value, end-value*, and *repeat-value* are independent of the specific input data, i.e., are constants or functions of the *number* of input items, n, but not of their specific values.

We allow each processor a local register (accumulator), ACC, for conditional testing and storage of partial results. The instruction $inst_x$ instantiated by the control program at processor x is a *numbered* sequence of basic operations from the set of primitives listed in Table 4.4.

Table 4.4. Processor instruction set

Symbol	Name	Meaning
$LDC\ x$	Load Constant	Place the value x in ACC
$LDA\ x$	Load (direct address)	$ACC := M[x]$
$LDI\ x$	Load (indirect address)	$ACC := M[M[x]]$
$STA\ x$	Store in memory (direct)	$M[x] := ACC$; ACC is set to 0.
$STI\ x$	Store in memory (indirect)	$M[M[x]] := ACC$; ACC is set to 0
$JUMP\ k$	Unconditional jump	Jump to instruction k
$JLT\ x,k$	Conditional jump	If $M[x] < 0$ then jump to instruction k
$JLE\ x,k$	Conditional jump	If $M[x] \leq 0$ then jump to instruction k
$JGT\ x,k$	Conditional jump	If $M[x] > 0$ then jump to instruction k
$JGE\ x,k$	Conditional jump	If $M[x] \geq 0$ then jump to instruction k
$JEQ\ x,k$	Conditional jump	If $M[x] = 0$ then jump to instruction k
$JNE\ x,k$	Conditional jump	If $M[x] \neq 0$ then jump to instruction k
$ADD\ x$	Addition	$ACC := ACC + M[x]$
$SUB\ x$	Subtraction	$ACC := ACC - M[x]$
$MULT\ x$	Multiplication	$ACC := ACC * M[x]$
$DIV\ x$	Integer Division	$ACC := ACC/M[x]$
$AND\ x$	Bitwise logical conjunction	$ACC := ACC \wedge M[x]$
NEG	Bitwise logical negation	$ACC := \neg ACC$
$HALT$	Halt	Stop execution

We now consider the parallel complexity measures used within this model. Let A be a P-RAM algorithm to compute some function $f(t_1, \ldots, t_n)$ employing instructions from Table 4.4. The *parallel runtime* of A on $< t_1, \ldots, t_n >$, denoted $T(t_1, \ldots, t_n)$, is the number of calls of **par do** multiplied by the worst case number of basic operations executed by a single processor during any of these calls. This is, of course, typically an overestimate of the actual parallel runtime; however, for the specific class of parallel algorithms with which we are concerned this is not significant.

The parallel runtime of A on inputs of size n, denoted $T(n)$, is $T(n) = \max_{\{t_1, \ldots, t_n\}} T(t_1, \ldots, t_n)$. The *processor requirements* of A on $< t_1, \ldots, t_n >$, denoted by $P(t_1, \ldots, t_n)$, is the maximum size of $| L |$ in an instruction **for** $x \in L$ **par do** ... that is executed by A. The processor requirements in inputs of size n, denoted $P(n)$, is

$$P(n) = \max_{\{t_1, \ldots, t_n\}} P(t_1, \ldots, t_n).$$

Finally, the *space* required by A on $< t_1, \ldots, t_n >$, denoted by $S(t1, \ldots, t_n)$, is the maximum value, s, such that $M[s]$ is referenced during the computation of A. The space required by A on inputs of size n, denoted $S(n)$, is $S(n) = \max_{\{t_1, \ldots, t_n\}} S(t_1, \ldots, t_n)$. It is assumed that $S(n) \geq P(n)$, i.e., that there is a memory location that is associated with each processor activated.

Let $f : \mathbf{N} \to \mathbf{N}$ be some function over $\mathbf{N}$. We say that a function $Q : \mathbf{N} \to \mathbf{N}$ has parallel-time (space, processor) complexity $f(n)$ is there exists a CREW P-RAM algorithm computing Q and such that $T(n) \leq f(n)(S(n) \leq f(n), P(n) \leq f(n))$.

We are particularly concerned with the complexity class of problems that can be solved with parallel algorithms with a polynomial number of processors and having polylogarithmic parallel runtime, i.e., the class of functions M for which there is an algorithm A achieving $T(n) \leq (\log n)^k$ and $P(n) \leq n^r$ for some constants k and r; these form the complexity class NC of problems regarded as having efficient parallel algorithms.

4.10 The Translation Process

We now provide a description of the translation from CREW P-RAM algorithm to DNA. The input to the process is a CREW P-RAM program A, which takes n input items, uses $S(n)$ memory locations, employs $P(n)$ different processors, and has a total running time of $T(n)$. The final output takes the form of a DNA algorithm Q with the following characteristics: Q takes as input an encoding of the initial input to A, returns as its output an encoding of the final state of the $S(n)$ memory locations after the execution of A, has a total running time of $O(T(n) \log S(n))$, and requires a total volume of DNA $O(T(n)P(n)S(n) \log S(n))$. The translation involves the following stages:

1. Translate the control program into a straight-line program
2. For each parallel processing instruction, construct a combinational circuit to simulate it
3. Cascade the circuits generated in Stage 2
4. Translate the resulting circuit into a DNA algorithm

We first translate the control program to a straight-line program SA. SA consists of a sequence of $R(n) \leq T(n)$ instructions, $I_k (1 \leq k \leq R(n))$, each of which is of the form **for** $x \in L_k$ **par do** $inst_k(x)$. We "unwind" the control program by replacing all loops involving m repetitions of an instruction sequence I by m copies $I_1, \ldots, I_m$ of I in sequence. Since the loops in the control program are dependent only on the number of inputs, this transformation can always be carried out from just the value of n.

Recall that $inst_k(x)$, the program instantiated on processor P_x by instruction A_k of the (modified) control program, is a sequence of numbered instructions drawn from Table 4.4. Furthermore, $inst_k(x)$ and $inst_k(y)$ are identical except for specific references to memory locations. It follows that in describing the translation to a combinational circuit, it suffices to describe the process for a single processor: an identical translation will then apply for all other processors working in parallel.

The simulation of the program running on a single processor lies at the core of the combinational circuit translation, and requires some initial transformation of the given program. To assist with this we employ the concept of a *program control graph*.

Definition 2. *Let $P = <p_1, p_2, \ldots, p_t >$ be an ordered sequence of numbered instructions from the set described in Table 4.4. The control graph, $G(V, E)$ of P is the directed graph with vertex set $\{1, 2, \ldots, t\}$ and edge set defined as follows: there is an edge (v, w) directed from v to w if instruction p_w immediately follows instruction p_v (note, "immediately follows" means $w = v + 1$ and p_v is not an unconditional jump) or p_v has the form $JUMP\ w$ or $JEQ\ x, w$. Notice that each vertex in G has either exactly one edge directed out of it or, if the vertex corresponds to a conditional jump, exactly two edges directed out of it. In the latter case, each vertex on a directed path from the source of the branch corresponds to an instruction which would be executed only if the condition determining the branch indicates so.*

We impose one restriction on programs concerning the structure of their corresponding control graphs.

Definition 3. *Let $G(V, E)$ be the control graph of a program P. G is well formed if, for any pair of directed cycles $C = < c_1, \ldots, c_r >$ and $D = < d_1, \ldots, d_s >$ in G, it holds that $\{c_1, \ldots, c_r\} \subset \{d_1, \ldots, d_s\}$, $\{d_1, \ldots, d_s\} \subset \{c_1, \ldots, c_r\}$, or $\{c_1, \ldots, c_r\} \cap \{d_1, \ldots, d_s\} = \emptyset$.*

It is required of the control graphs resulting from programs that they be well formed. Notice that the main property of well-formed graphs is that any two loop structures are either completely disjoint or properly nested. The next stage is to transform the control graph into a directed acyclic graph which represents the same computation as the original graph. This can be accomplished by unwinding loop structures in the same way as was done for the main control program.

Let H_P be the control graph that results after this process has been completed. Clearly H_P is acyclic. Furthermore, if the program corresponding to H_P is executed, the changes made to the memory, M, are identical to those that would be made by P. Finally, the number of vertices (corresponding to the number of instructions in the equivalent program) is $O(t)$, where t is the worst case number of steps executed by P.

The final stage is to reduce H_P to a branch-free graph. Since cycles have already been eliminated this involves merging conditional branches into a sequence of instructions. This is accomplished by the transformation illustrated in Fig. 4.9. In this process both branches of the conditional are executed; however, only the branch for which the condition is true will affect the contents of memory locations.

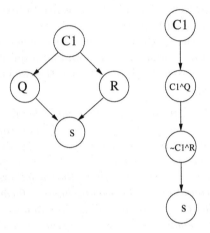

Fig. 4.9. Merging of conditional branches

If G_P is the initial control graph of P and L_P is the graph resulting after all of the transformations above have been effected, then L_P is called the *linear control graph* of the program P.

The total effect of the transformation described above is to replace the initial program by a sequence of instructions $< Q_1, Q_2, \ldots, Q_m >$, where m is at most a constant factor larger than the runtime of P. Furthermore, each instruction Q_i has the form

$$< condition >< active - instruction >$$

where $< condition >$ is a composition of a constant number of tests arising from JEQ instructions, and $< active - instruction >$ is any non-branching instruction. The interpretation of this form is that $< active - instruction >$ changes the contents of a memory location if and only if $< condition >$ is true.

We then convert the linear control graph into a series of combinational circuits. For each of the $R(n)$ parallel processing (i.e., **par do**) instructions in SA we construct a combinational circuit C_k with the following specification: C_k takes as input $r \times S(n)$ Boolean values arranged in $S(n)$ blocks of r bits (we assume, without loss of generality, that $r \leq S(n)$); the values in the ith

block $B_{k-1}(i)$ correspond to the contents of memory location $M[i]$ prior to the execution of I_k by SA; C_k produces as its *output* $r \times S(n)$ Boolean values arranged as $S(n)$ r-bit blocks, the values in the ith block $B_k(i)$ corresponding to the contents of memory location $M[i]$ after the execution of I_k by SA.

As a consequence of the translation just described, the programs corresponding to the actions of each processor working in parallel at some stage of the computation consist of *identical* (except for memory references) linear control graphs. A typical linear graph consists of processing some values stored in memory (using the accumulator), storing a value computed in memory, and repeating this sequence until its computation is completed. We can thus simulate this process by building a circuit which consists of several levels –each level mimics the calculations performed in the program and ends with either a halt instruction or a store instruction. What has to be ensured in the circuit simulation is that the store and halt instructions have an effect only if their controlling conditions evaluate to true. Thus, let $L = < l_1, l_2, \ldots, l_t >$ be a sequence of instructions such that l_t has the form $< conditional > HALT$, $< conditional > STA$, or $< conditional > STI$, and $l_k (1 \le k < t)$ is a load or arithmetic instruction.

We now describe in detail the combinational circuit, CL, described above, which carries out the following: CL has $S(n) \times r$ input bits that represent the current contents of the $S(n)$ r-bit words in the memory M; CL outputs $S(n) \times r$ bits that represent the contents of M after $l_1, \ldots, l_t$ have executed. The output differs from the input only if the conditional qualifying l_t evaluates to true. In addition, CL employs the following Boolean variables which are instantiated from the specific program instructions:

- *cond*: a single Boolean value which will compute to 1 if and only if the conditional controlling l_t is true
- $ADDR[0..m-1]$: an m-bit representation of the address of a location in memory (so that $m = \log_2 S(n)$).

We employ the following shorthand notations:

- $bin(k)$ denotes the m-bit representation of k $(0 \le k < S(n))$
- If x is a single Boolean value and $R[] = < r_0, \ldots, r_k >$, an ordered set of Boolean values, the notation $x < op > R[]$ (where op is a Binary boolean operator) denotes the set of values $R[x < op > r_0, \ldots, x < op > r_k]$
- $M_x^{in}[]$ denotes the r-bit content of $M[x]$ before the start of a computation (i.e., at input) and $M_x^{out}[]$ denotes the contents of the same location after the computation has finished.

In order to select data from memory we employ a module

$$FETCH : \{0,1\}^{r \times S(n)} \times \{0,1\}^m \to \{0,1\}^r$$

$FETCH$ takes as input the repreentation of M and an m-bit address, α, and outputs the r bits in M_α^{in}. The Boolean realization of $FETCH(M(0, \ldots, S(n-$

1)), $ADDR[])$ is given by

$$FETCH[0...r-1] = \overset{S(n)-1}{\underset{k=0}{\vee}} M_k[] \wedge (bin(k) \equiv ADDR[])$$

The total number of two-input gates required to realize this operation is

$$Size(FETCH) \leq S(n) - 1 + S(n)(r+m) = O(S(n) \log S(n))$$
$$Depth(FETCH) \leq m + \log(m) = O(\log S(n)))$$

We now describe in detail the combinational circuit associated with each operation in Table 4.4.

Load constant x
    This merely involves transcribing the r-bit binary representation of the constant x onto the r-wires representing the accumulator contents.
$Size(LDC) = 0$; $Depth(LDC) = 0$

Load direct x
    This is equivalent to $FETCH(M[], x)$ and transcribing the output to the accumulator wires.
$Size(LDA) = O(S(n) \log S(n))$; $Depth(LDA) = O(\log S(n))$

Load indirect x
    Indirect addressing involves two successive invocations of the $FETCH$ module, i.e., $FETCH(M[], FETCH[M[], x))$ and transcribing to the accumulator.
$Size(LDI) = O(S(n) \log S(n))$; $Depth(LDI) = O(\log S(n))$

ADD x
    $ADD(ACC[], FETCH(M[], x))$. An addition module, $ADD(x, y)$ adds two r-bit values and returns an r-bit value (overflow is assumed to cause an error). A number of combinational circuits exist which achieve this in $O(r)$ gates and $O(\log r)$ depth, e.g., [91]. Hence, the dominant complexity contribution arises from fetching the operand. In total we have
$Size(ADD) = O(S(n) \log S(n))$; $Depth(ADD) = O(\log S(n))$

SUB x
    Similar to ADD using, for example, complement arithmetic.

MULT x

$MULT(ACC[], FETCH(M[], x))$. A multiplication module, $MULT(x, y)$ takes two r-bit values and computes their produce (again, overflow is an error). The technique described by Schonhage and Strassen [140] yields a combinational circuit of size $O(r \log r \log \log r)$ and depth $O(\log r)$. Hence, the entire operation, including the $FETCH$ stage requires
$Size(MULT) = O(S(n) \log S(n)); \ Depth(MULT) = O(\log S(n))$

DIV x

$DIV(ACC[], FETCH(M[], x))$. Beame, Cook, and Hoover [26] describe a division network which, when applied to r-bit operands has size $O(r^4 \log^3 r)$ and depth $O(\log r)$. Thus,
$Size(DIV) = O(S(n) \log S(n) + r^4 \log^3 r); \ Depth(DIV) = O(\log S(n))$

AND x

$AND(ACC[], FETCH(M[], x))$. The bitwise conjunction requires just r gates and can be accomplished in depth 1. Thus,
$Size(AND) = O(S(n) \log S(n)); \ Depth(AND) = O(\log S(n))$

NEG

No references to memory are required, since this operation is performed directly on the accumulator.
$Size(NEG) = r; \ Depth(NEG) = 1$

STA x

Copy the m-bit value of x onto $ADDR[0, \ldots, m-1]$; then for each memory location y, $0 \le y < S(n)$, compute the following:
$M_y^{out} = (\overline{cond} \vee \overline{(bin(y) \equiv ADD[])}) \wedge M_y^{in} \vee (cond \wedge (bin(y) \equiv ADDR[])) \wedge ACC[]$
For brevity we denote this function by $STORE(ACC[], cond, ADDR[])$. $STORE$ is a function
$STORE : \{0,1\}^{r \times S(n)} \times \{0,1\}^r \times \{0,1\} \times \{0,1\}^m \rightarrow \{0,1\}^{r \times S(n)}$
Thus, a memory location is changed only if the condition governing the instruction is true and the location specified matches the output address. The condition governing the instruction involves only a constant number of tests of particular memory locations and can therefore be computed in $O(S(n) \log S(n))$ gates and $O(\log S(n))$ depth. We note that the update of all $S(n)$ memory locations can be carried out in parallel. Thus,
$Size(STA) = S(n)((1 + m + m) + (1 + m + m) + 1) = O(S(n) \log S(n));$
$Depth(STA) = O(\log S(n))$

STI x

This is identical to STA except that calculating $ADDR[] = FETCH(M[], x)$ is performed to determine the actual memory location to be used. In total, this gives

$Size(STI) = O(S(n) \log S(n)); \ Depth(STI) = O(\log S(n))$

We then cascade the $R(n)$ combinational circuits produced in the preceding construction stage (Fig. 4.10) and transform this resulting circuit into one over the basis $NAND$. This completes the description of the translation of the program of instructions to a combinational circuit.
In total, we have,

Theorem 1. *Let A be a CREW P-RAM algorithm using $P(n)$ processors and $S(n)$ memory locations (where $S(n) \geq P(n)$) and taking $T(n)$ parallel time. Then there is a combinational logic circuit, C, computing exactly the same function as A and satisfying*

$$Size(C) = O(T(n)P(n)S(n) \log S(n)); \ Depth(C) = O(T(n) \log S(n)).$$

Proof. The modification of the control program allows A to be simulated by at most $R(n)$ blocks which simulate each parallel instantiation (c.f. Fig. 4.10). In one block there are at most $P(n)$ different circuits that update the $S(n)$ locations in the global memory. Each such circuit is a translation of a program comprising $O(T_k(n))$ instructions, where $T_k(n)$ is the runtime of the program run on each processor during the kth cycle of the control program. Letting $C_k(n)$ denote the circuit simulating a program in the kth cycle, we have

$$Depth(C) \leq \sum_{k=1}^{R(n)} Depth(C_k(n)) \leq O(\log S(n)) \sum_{k=1}^{R(n)} O(T_k(n))$$

$$= O(T(n) \log S(n))$$

For the circuit size analysis,

$$Size(C) \leq P(n) \sum_{k=1}^{R(n)} Size(C_k(n)) \leq O(P(n)T(n)S(n) \log S(n)). \ \square$$

Corollary 1: If A is an NC algorithm, i.e., requires $O(n^k)$ processors and $O(\log^r n)$ time, then the circuit generated by the translation above has polynomially many gates and polylogarithmic depth.
Proof: Immediate from Theorem 1, given that $S(n)$ must be polynomial in n.
$\square$

4.11 Assessment

So far we have presented a detailed, full-scale translation from CREW P-RAM algorithms to operations on DNA. The volume of DNA required and the running time of the DNA realization are within a factor $\log S$ of the

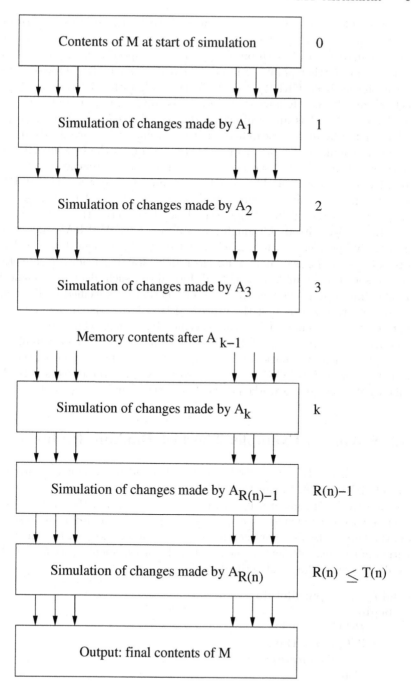

Fig. 4.10. Combinational circuit block simulation of control program

best attainable (where S is the total space used by the P-RAM algorithm). One consequence of this is that the class, NC, of efficient parallel algorithms can be realized in DNA with only a small loss of speed and increase in size. We note some further points of interest about our simulation process. First, the translation from CREW P-RAM to DNA algorithm is an effective and practical one. It thus follows that, in at least one computational paradigm, the CREW P-RAM is an architecture that *can* be realized. This is despite the fact, noted by many researchers, that the technological overheads implied by a global common memory regime, to say nothing of the Concurrent Read facility, which means that within traditional silicon approaches the CREW P-RAM is not a constructible, scalable machine. In other words, our DNA realization gives a feasible, concrete implementation of what has been regarded as a purely abstract device. The second point concerns the nature of the translation process itself. Although no published CREW P-RAM algorithms are actually specified at the "machine code" level we regard as the basic processor instruction set, it is quite clear that, starting from a (sequential) program specified in an *ersatz*-ALGOL formalism, such algorithms could be compiled into a program over the basic instruction set specified in Table 4.4. It is also clear, however, that the process of translating the "assembly level" program to a combinational circuit specification is a mechanical one, and, furthermore, the creation of DNA strands from a given $NAND$ circuit can also be performed automatically. If we combine these chains of ideas, then we see that a $high - level$ CREW P-RAM to DNA *compiler* process is readily available. We now give a worked example of this compilation process.

4.12 A Worked Example: The List Ranking Problem

We give, in this section, a worked example of the ideas presented above on a typical CREW P-RAM algorithm. The example we choose is the list ranking algorithm from [64]: given a list, L, of $n = 2^m$ elements, the problem is to associate with each element k a value $rank(k)$ corresponding to its distance from the end of the list. For each list element k a location $next(k)$ gives the location of the next element in the list. For the last element, $next(k) = k$. We describe the algorithm as described in [64]:

for all $k \in L$ **in parallel do**
    begin
        $P(k) := next(k);$
        if $P(k) \neq k$ **then**
            $distance(k) := 1;$
        else
            $distance(k) := 0;$
    end
repeat $\log n$ **times**

for all $k \in L$ **in parallel do**
    if $P(k) \neq P(P(k))$ **then**
    begin
        $distance(k) := distance(k) + distance(P(k));$
        $P(k) := P(P(k));$
    end;
for all $k \in L$ **in parallel do**
    $rank(k) := distance(k);$

Obviously this solves the list ranking problem in $O(\log n)$ parallel steps, using n processors and $O(n)$ memory locations. We consider the realization of this for $n = 2^m$. In the translation, a total of $5n$ memory locations are used, which are laid out as follows:

$M[1], \ldots, M[n]$ hold the list pointers $next(k)(1 \leq k \leq n)$
$M[n+1], \ldots, M[2n]$ hold the values $P(k)$
$M[2n+1], \ldots, M[3n]$ hold $distance(k)$
$M[3n+1], \ldots, M[4n]$ are used for temporary results
$M[4n+1], \ldots, M[5n]$ hold the output values $rank(k)$

The unwound control program contains precisely $m + 2$ **in parallel do** instructions, the middle m of which are identical. For the first instruction we have

for all $k \in L$ **in parallel do**
    begin
        $P(k) := next(k);$
        if $P(k) \neq k$ **then**
            $distance(k) := 1;$
        else
            $distance(k) := 0;$
    end

In terms of the basic instruction set this translates to

```
1.   LDA k
2.   STA n + k          /* P(k) := next(k)
3.   LDC k
4.   SUB n + k
5.   STA 3n + k
6.   JEQ 3n + k, 10     /* P(k) ≠ k
7.   LDC 1
8.   STA 2n + k         /* distance(k) := 1
9.   JUMP 12
```

10. LDC 0
11. STA $2n + k$ /* $distance(k) := 0$
12. HALT

(Note that the memory addresses given translate to fixed values for a fixed n and processor identifier k).

The second (block of) m instructions follows:

> **for all** $k \in L$ **in parallel do**
>   **if** $P(k) \neq P(P(k))$ **then**
>   **begin**
> $distance(k) := distance(k) + distance(P(k))$;
> $P(k) := P(P(k))$;
>   **end;**

1. LDC n /* Base for $P(k)$
2. ADD $n + k$ /* $n + P(k)$, i.e. address of $P(P(k))$
3. STA $3n + k$
4. LDI $3n + k$ /* $P(P(k))$
5. SUB $n + k$
6. STA $3n + k$
7. JEQ $3n + k, 19$ /* $P(k) \neq P(P(k))$
8. LDC $2n$ /* Base for $distance(k)$
9. ADD $n + k$ /* $2n + P(k)$, i.e. address of $distance(P(k))$
10. STA $3n + k$
11. LDI $3n + k$
12. ADD $2n + k$ /* $distance(P(k)) + distance(k)$
13. STA $2n + k$ /* $distance(k) := distance(k) + distance(P(k))$
14. LDC n /* Base for $P(k)$
15. ADD $n + k$ /* $n + P(k)$, i.e. address of $P(P(k))$
16. STA $3n + k$
17. LDI $3n + k$
18. STA $n + k$ /* $P(k) := P(P(k))$
19. HALT

Finally,

> **for all** $k \in L$ **in parallel do**
> $rank(k) := distance(k)$;

1.   LDA $2n + k$
2.   STA $4n + k$ /* $rank(k) := distance(k)$
3.   HALT

Combinational circuit realization

First instruction block
Input: $M[1], \ldots, M[5n]$, each consisting of r bits
Output: $M[1], \ldots, M[5n]$ representing the contents of memory after the n processors have executed the first instruction.

For each processor k the following combinational circuit is used:

$ACC[] := FETCH(M[\cdots], bin(k));$
$M[\cdots] := STORE(ACC[], 1, bin(n + k));$
$ACC[] := bin(k);$
$ACC[] := SUB(ACC[], FETCH(M[\cdots], bin(n + k)));$
$M[\cdots] := STORE(ACC[], 1, bin(3n + k));$
$ACC := FETCH(M[\cdots], bin(3n + k));$
$cond := \overset{r-1}{\underset{i=0}{\wedge}} (ACC[i] \equiv 0);$
$ACC[] := bin(1);$
$M[\cdots] := STORE(ACC[], cond, bin(2n + k));$
$ACC[] := bin(0);$
$M[\cdots] := STORE(ACC[], \neg cond, bin(2n + k));$

Total size of first instruction $= n \times (3Size(FETCH) + 4Size(STORE) + Size(SUB) + Size(cond - eval)) = O(n^2 \log n)$.
Depth is at most $3Depth(FETCH) + 4Depth(STORE) + Depth(SUB) + Depth(cond - eval) = O(\log n)$.

Second instruction block
Input: $M[1], \ldots, M[5n]$, each consisting of r bits
Output: $M[1], \ldots, M[5n]$ representing the contents of memory after the n processors have executed the first instruction.

$ACC[] := bin(n)$
$ACC[] := ADD(ACC[], FETCH(M[\cdots], bin(n + k)));$
$M[\cdots] := STORE(ACC[], 1, bin(3n + k));$
$ACC[] := FETCH(M[\cdots], FETCH(M[\cdots], bin(3n + k)));$
$ACC[] := SUB(ACC[], FETCH(M[\cdots], bin(n + k)));$
$M[\cdots] := STORE(ACC[], 1, bin(3n + k));$
$ACC[] := FETCH(M[\cdots], bin(3n + k));$
$cond := \neg(\overset{r-1}{\underset{i=0}{\wedge}} (ACC[i] \equiv 0));$

$ACC[] := bin(2n);$
$ACC[] := ADD(ACC[], FETCH(M[\cdots], bin(n + k)));$
$M[\cdots] := STORE(ACC[], cond, bin(3n + k));$
$ACC[] := FETCH(M[\cdots], FETCH(M[\cdots], bin(3n + k)));$
$ACC[] := ADD(ACC[], FETCH(M[\cdots], bin(2n + k)));$
$M[\cdots] := STORE(ACC[], cond, bin(2n + k));$
$ACC[] := bin(n);$
$ACC[] := ADD(ACC[], FETCH(M[\cdots], bin(n + k)));$
$M[\cdots] := STORE(ACC[], cond, bin(3n + k));$
$ACC[] := FETCH(M[\cdots], FETCH(M[\cdots], bin(3n + k)));$
$M[\cdots] := STORE(ACC[], cond, bin(n + k));$

Total size $= n \times (12Size(FETCH) + 6Size(STORE) + 4Size(ADD) + Size(SUB) + Size(cond - eval)) = O(n^2 \log n).$
Depth is at most $(12Depth(FETCH) + 6Depth(STORE) + 4Depth(ADD) + Depth(SUB) + Depth(cond)) = O(\log n).$

This instruction, however, is repeated $\log n$ times, giving a total size of $O((n \log n)^2)$ and depth $O(log^2 n)$.

Final instruction
Input: $M[1], \ldots, M[5n]$, each consisting of r bits
Output: $M[1], \ldots, M[5n]$ representing the contents of memory after the n processors have executed the first instruction.

$ACC[] := FETCH(M[\cdots], bin(2n + k));$
$M[\cdots] := STORE(ACC[], 1, bin(4n + k));$

Total size $= n \times (Size(FETCH) + Size(STORE)) = O(n^2 \log n).$ Total depth $= Depth(FETCH) + Depth(STORE) = O(\log n).$

In total we have a combinational circuit for list ranking which has size $O((n \log n)^2)$ and depth $O(log^2 n)$.

4.13 Summary

In this chapter we have emphasized the rôle that complexity considerations are likely to play in the identification of "killer applications" for DNA computation. We have examined how time complexities have been estimated within the literature. We have shown that these are often likely to be inadequate from a realistic point of view. In particular, many authors implicitly assume that arbitrarily large numbers of laboratory assistants or robotic arms are available for the mechanical handling of tubes of DNA. This has often led to serious underestimates of the resources required to complete a computation.

We have proposed a so-called *strong* model of DNA computation, which we believe allows *realistic* assessment of the time complexities of algorithms within it. This model, if the *splice* operation is trivially included, not only provides realistic estimates of time complexities, but is also Turing-complete. We have also demonstrated how existing models of computation (Boolean circuits and the P-RAM) may be effectively simulated in DNA.

We believe that success in the search for "killer applications" is the only means by which there will be sustained practical interest in DNA computation. Success is only a likely outcome if DNA computations can be described that will require computational resources of similar magnitude to those required by conventional solutions. If, for example, we were to establish polylogarithmic time computations using only a polynomial volume of DNA, then this would be one scenario in which "killer applications" might well ensue. In this case, we might imagine that the vast potential for parallelisation may finally be effectively harnessed.

4.14 Bibliographical Notes

For an in-depth yet accessible treatment of computational complexity, the reader is directed to [145]. This text includes short sections on the complexity class NC and P-completeness. Detailed discussion of parallel computation (with specific reference to the P-RAM) is collected in [66].

5

Physical Implementations

"No amount of experimentation can ever prove me right; a single experiment can prove me wrong." – Albert Einstein

5.1 Introduction

This chapter provides an introduction to the implementation of DNA computations. We concentrate in particular on a full description of two filtering models (Adleman's and parallel filtering). We highlight the practical implementation problems inherent in all models, and suggest possible ways to alleviate these. We also describe other key successful experimental implementations of DNA-based computations.

5.2 Implementation of Basic Logical Elements

In 1982, Bennett [29] proposed the concept of a "Brownian computer" based around the principle of reactant molecules touching, reacting, and effecting state transitions due to their random Brownian motion. Bennett developed this idea by suggesting that a Brownian Turing Machine could be built from a macromolecule such as RNA. "Hypothetical enzymes", one for each transition rule, catalyze reactions between the RNA and chemicals in its environment, transforming the RNA into its logical successor.

Conrad and Liberman developed this idea further in [46], in which the authors describe parallels between physical and computational processes (for example, biochemical reactions being employed to implement basic switching circuits). They introduce the concept of molecular level "word processing" by describing it in terms of transcription and translation of DNA, RNA processing, and genetic regulation. However, the paper lacks a detailed description of the biological mechanisms highlighted and their relationship with "traditional" computing. As the authors themselves acknowledge, "our aspiration is not to provide definitive answers ... but rather to show that a number of seemingly disparate questions must be connected to each other in a fundamental way." [46]

In [45], Conrad expanded on this work, showing how the information processing capabilities of organic molecules may, in theory, be used in place of digital switching components (Fig. 5.1a). Enzymes may *cleave* specific substrates by severing covalent bonds within the target molecule. For example, as we have seen, restriction endonucleases cleave strands of DNA at specific points known as *restriction sites*. In doing so, the enzyme switches the *state* of the substrate from one to another. Before this process can occur, a recognition process must take place, where the enzyme distinguishes the substrate from other, possibly similar molecules. This is achieved by virtue of what Conrad refers to as the "lock-key" mechanism, whereby the complementary structures of the enzyme and substrate fit together and the two molecules bind strongly (Fig. 5.1b). This process may, in turn, be affected by the presence or absence of *ligands*. Allosteric enzymes can exist in more than one conformation (or "state"), depending on the presence or absence of a ligand. Therefore, in addition to the active site of an allosteric enzyme (the site where the substrate reaction takes place) there exists a ligand binding site which, when occupied, changes the conformation and hence the properties of the enzyme. This gives an additional degree of control over the switching behavior of the entire molecular complex.

In [17], Arkin and Ross show how various logic gates may be constructed using the computational properties of enzymatic reaction mechanisms (also see [34] for a review of this work). In [34], Bray also describes work [79, 80] showing how chemical "neurons" may be constructed to form the building blocks of logic gates.

5.3 Initial Set Construction Within Filtering Models

All filtering models use the same basic method for generating the initial set of strands. An essential difficulty in all filtering models is that initial multi-sets generally have a cardinality which is exponential in the problem size. It would be too costly in time, therefore, to generate these individually. What we do in practice is to construct an initial solution, or *tube*, containing a polynomial number of distinct strands. The design of these strands ensures that the exponentially large initial multi-sets of our model will be generated automatically. The following paragraph describes this process in detail.

Consider an initial set of all elements of the form $p_1k_1, p_2k_2, \ldots, p_nk_n$. This may be constructed as follows. We generate an oligonucleotide (commonly abbreviated to *oligo*) uniquely encoding each possible subsequence p_ik_i where $1 \leq i \leq n$ and $1 \leq k_i \leq k$. Embedded within the sequence representing p_i is our chosen restriction site. There are thus a polynomial number, nk, of distinct oligos of this form. The task now is how to combine these to form the desired

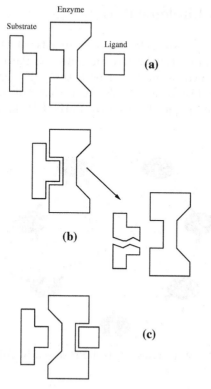

Fig. 5.1. (a) Components of enzymatic switch. (b) Enzyme recognizes substrate and cleaves it. (c) Ligand binds to enzyme, changing its conformation; enzyme no longer recognizes substrate

initial multi-set. This is achieved as follows: for each pair $(p_i k_i, p_{i+1} k_{i+1})$ we construct an oligo which is the concatenation of the complement of the second half of the oligo representing $p_i k_i$ and the complement of the first half of the oligo representing $p_{i+1} k_{i+1}$. We also construct oligos that are the complement of the first half of the oligo representing $p_1 k_1$ and the last half of the oligo representing $p_n k_n$. There is therefore a total of $2nk + 1$ oligos in solution.

The effect of adding these new oligos is that double-stranded DNA will be formed in the tube and one strand in each will be an element of the desired initial set. The new oligos have, through annealing, acted as "splints" to join the first oligos in the desired sequences. These splints may then be removed from solution (assuming that they are biotinylated).

5.4 Adleman's Implementation

Adleman utilized the incredible storage capacity of DNA to implement a brute-force algorithm for the directed Hamiltonian Path Problem (HPP). Recall that the HPP involves finding a path through a graph that visits each vertex exactly once. The instance of the HPP that Adleman solved is depicted in Fig. 5.2, with the unique Hamiltonian Path (HP) highlighted by a dashed line.

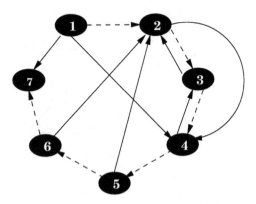

Fig. 5.2. Instance of the HPP solved by Adleman

Adleman's approach was simple:

1. Generate strands encoding random paths such that the Hamiltonian Path (HP) is represented with high probability. The quantities of DNA used far exceeded those necessary for the small graph under consideration, so it is likely that *many* strands encoding the HP were present.
2. Remove all strands that do not encode the HP.
3. Check that the remaining strands encode a solution to the HPP.

The individual steps were implemented as follows:

Stage 1: Each vertex and edge was assigned a distinct 20-mer sequence of DNA (Fig. 5.3a). This implies that strands encoding a HP were of length 140 b.p. Sequences representing edges act as 'splints' between strands representing their endpoints (Fig. 5.3b).

In formal terms, the sequence associated with an edge $i \rightarrow j$ is the 3' 10-mer of the sequence representing v_i followed by the 5' 10-mer of the sequence representing v_j. These oligonucleotides were then combined to form strands encoding random paths through the graph. An (illegal) example path ($v_1 \rightarrow v_2 \rightarrow v_3 \rightarrow v_4$) is depicted in Fig. 5.4.

Fixed amounts (50 pmol) of each oligonucleotide were mixed together in a single ligation reaction. At the end of this reaction, it is assumed that a

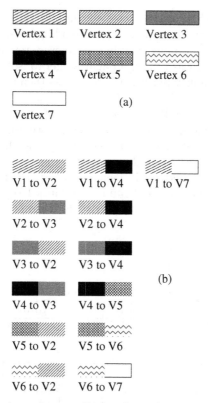

(a)

(b)

Fig. 5.3. Adleman's scheme for encoding paths - schematic representation of oligos

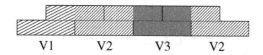

Fig. 5.4. Example path created in Adleman's scheme

strand representing the HP is present with high probability. This approach
solves the problem of generating an exponential number of different paths
using a polynomial number of initial oligonucleotides.

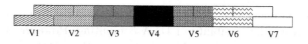

Fig. 5.5. Unique Hamiltonian path

Stage 2: PCR was first used to massively amplify the population of oligonucleotides encoding paths starting at v_1 and ending at v_7. Next, strands that do not encode paths containing exactly n visits were removed. The product of the PCR amplification was run on an agarose gel to isolate strands of length 140 b.p. A series of affinity purification steps was then used to isolate strands encoding paths that visited each vertex exactly once.

Stage 3: Graduated PCR was used to identify the unique HP that this problem instance provides. For an n-vertex graph, we run $n-1$ PCR reactions, with the strand representing v_1 as the left primer and the complement of the strand representing v_i as the right primer in the ith lane. The presence of molecules encoding the unique HP depicted in Fig. 5.2 should produce bands of length 40, 60, 80, 100, 120, and 140 b.p. in lanes 1 through 6, respectively. This is exactly what Adleman observed. The graduated PCR approach is depicted in Fig. 5.6.

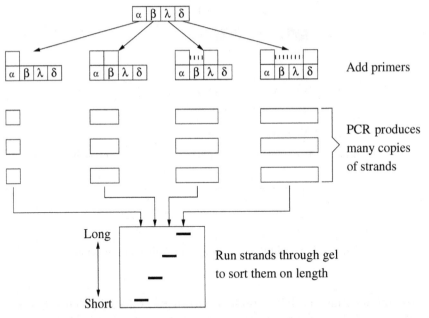

Fig. 5.6. Graduated PCR

Adleman's experiment was remarkable in that it was the first to demonstrate in the laboratory the feasibility of DNA computing. However, we note that it was performed on a single problem instance with just one HP. No control experiments were performed for cases without Hamiltonian Paths. The final detection step is problematic, due to reliance on the error prone PCR procedure. In addition, the use of affinity purification is also error prone, which may mean that the experiment will not successfully scale up. We consider these issues in later sections.

5.5 Evaluation of Adleman's Implementation

We describe later how the various multi-set operations described in the previous section may be realized thorough standard DNA manipulation techniques. However, it is convenient at this point to emphasize two impediments to effective computation by this means. The first hampers the problem size that might be effectively handled, and the second casts doubt on the potential for biochemical success of the precise implementations that have been proposed.

Naturally, the strings making up the multi-sets are encoded in strands of DNA in all the proposed implementations. Consider for a moment what volume of DNA would be required for a typical NP-complete problem. The algorithms mentioned earlier require just a polynomial number of DNA manipulation steps. For the NP-complete problems there is an immediate implication that an exponential number of parallel operations would be required within the computation. This in turn implies that the tube of DNA must contain a number of strands which is exponential in the problem size. Despite the molecular dimensions of the strands, for only moderate problem sizes (say, $n \sim 20$ for the Hamiltonian Path problem) the required volume of DNA would make the experiments impractical. As Hartmanis points out in [76], if Adleman's experiment were scaled up to 200 vertices the weight of DNA required would exceed that of the earth. Mac Dónaill also presents an analysis of the scalability of DNA computations in [53], as do Linial and Linial [97], Lo et al. [101], and Bunow [38].

We note that [19] has described DNA algorithms which reduce the problem just outlined; however, the "exponential curse" is inherent in the NP-complete problems. There is the hope, as yet unrealized (despite the claims of [24]) that for problems in the complexity class P (i.e. those which can be solved in sequential polynomial time) there may be DNA computations which only employ polynomial sized volumes of DNA.

We now consider the potential for biochemical success that was mentioned earlier. It is a common feature of *all the early proposed implementations* that the biological operations to be used are assumed to be error free. An operation central to and frequently employed in most models is the *extraction* of DNA strands containing a certain sequence (known as *removal by DNA hybridization*). The most important problem with this method is that it is not 100% specific,[1] and may at times inadvertently remove strands that *do not* contain the specified sequence. Adleman did not encounter problems with extraction because in his case only a few operations were required. However, for a large problem instance, the number of extractions required may run into hundreds,

---

[1] The actual specificity depends on the concentration of the reactants.

or even thousands. For example, a particular DNA-based algorithm may rely upon repeated "sifting" of a "soup" containing many strands, some encoding legal solutions to the given problem, but most encoding illegal ones. At each stage, we may wish to extract only strands that satisfy certain criteria (i.e., they contain a certain sequence). Only strands that satisfy the criteria at one stage go through to the next. At the end of the sifting process, we are hopefully left only with strands that encode legal solutions, since they satisfy all criteria. However, assuming 95% efficiency of the extraction process, after 100 extractions the probability of us being left with a soup containing (a) a strand encoding a legal solution and (b) no strands encoding illegal solutions is about 0.006. Repetitive extraction will not guarantee 100% efficiency, since it is impossible to achieve the conditions whereby only correct hybridization occurs. Furthermore, as the length of the DNA strands being used increases, so does the probability of incorrect hybridization.

These criticisms have been borne out by recent attempts [86] to repeat Adleman's experiment. The researchers performed Adleman's experiment twice; once on the original graph as a positive control, and again on a graph containing no Hamiltonian path as a negative control. The results obtained were inconclusive. The researchers state that *"at this time we have carried out every step of Adleman's experiment, but have not gotten an unambiguous final result."*

Although attempts have been made to reduce errors by (1) simulation of highly reliable purification using a sequence of imperfect operations [90] and (2) application of PCR at various stages of the computation [32], it is clear that reliance on affinity purification must be minimized or, ideally, removed entirely. In [12], we describe one possible error-resistant model of DNA computation that removes the need for affinity purification within the main body of the computation. It is proposed that affinity purification be replaced by a new enzymatic removal technique.

In [93], Kurtz et al. consider the effect of problem size on the initial concentrations of reactants and analyze the subsequent probability of a correct solution being produced. They claim that, without periodic amplification of the working solution, the concentration of strands drops exponentially with time to "homeopathic levels." One proposal to reduce strand loss during computations is described in [100]. Rather than allowing strands to float free in solution, the authors describe a surface-based approach, whereby strands are immobilized by attachment to a surface (glass is used in the experiments described in [100], although gold and silicon are other possible candidates). The attachment chemistry is described in detail in [72]. This model is similar to that described in [12], in that it involves selective destruction of specific strands, although in this case Exonuclease is used to destroy unmarked rather than marked strands. Preliminary experimental results suggest that strand loss is indeed reduced, although the scalability of this approach is questionable due to the two-dimensional nature of the surface.

5.6 Implementation of the Parallel Filtering Model

Here we describe how how the set operations within the Parallel Filtering Model described in Section 3.2 may be implemented.

Remove

$remove(U, \{S_i\})$ is implemented as a composite operation, comprised of the following:

- $mark(U, S)$. This operation marks all strings in the set U which contains at least one occurrence of the substring S.
- $destroy(U)$. This operation removes all marked strings from U.

$mark(U, S)$ is implemented by adding to U many copies of a primer corresponding to $\overline{S}$ (Fig. 5.7b). This primer only anneals to single strands containing the subsequence S. We then add DNA polymerase to extend the primers once they have annealed, making only the single strands containing S double stranded (Fig. 5.7b).

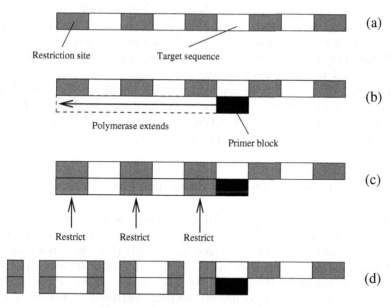

Fig. 5.7. Implementation of *destroy*

We may then *destroy* strands containing S by adding the appropriate restriction enzyme. Double-stranded DNA (i.e. strands marked as containing S) is cut at the restriction sites embedded within, single strands remaining intact

(Fig. 5.7c). We may then remove all intact strands by separating on length using gel electrophoresis. However, this is not strictly necessary, and leaving the fragmented strands in solution will not affect the operation of the algorithm.

Union

We may obtain the *union* of two or more tubes by simply mixing their contents together, forming a single tube.

Copy

We obtain i "copies" of the set U by splitting U into i tubes of equal volume. We assume that, since the initial tube contains multiple copies of each candidate strand, each tube will also contain many copies.

Select

We can easily detect remaining *homogeneous* DNA using PCR and then sequence strands to reveal the encoded solution to the given problem. One problem with this method is that there are often multiple correct solutions left in the soup which must be sequenced using nested PCR. This technique is only useful when the final solution is known in advance. Also, the use of PCR may introduce an unacceptable level of error in the read-out procedure. A possible solution is to use *cloning*.

Although the initial tube contains multiple copies of each strand, after many *remove* operations the volume of material may be depleted below an acceptable empirical level. This difficulty can be avoided by periodic amplification by PCR (this may also be performed after *copy* operations).

5.7 Advantages of Our Implementation

As we have shown, algorithms within our model perform successive "filtering" operations, keeping *good* strands (i.e., strands encoding a legal solution to the given problem) and destroying *bad* strands (i.e., those not doing so). As long as the operations work correctly, the final set of strands will consist only of good solutions. However, as we have already stated, errors can take place. If either good strands are accidentally destroyed or bad strands are left to survive through to the final set, the algorithm will fail. The main advantage of our model is that it doesn't repeatedly use the notoriously error prone separation by DNA hybridization method to extract strands containing a certain subsequence. Restriction enzymes are *guaranteed* [31, page 9][2]

---

[2] "New England Biolabs provides a color-coded 10X NEBuffer with each restriction endonuclease to ensure optimal (100%) activity."

to cut any double-stranded DNA containing the appropriate restriction site, whereas hybridization separation is never 100% efficient. Instead of extracting *most* strands containing a certain subsequence we simply destroy them with high probability, without harming those strands that *do not* contain the subsequence. In reality, even if restriction enzymes have a small nonzero error rate associated with them, we believe that it is far lower than that of hybridization separation. Another advantage of our model is that it minimizes physical manipulation of tubes during a computation. Biological operations such as pipetting, filtering, and extraction lose a certain amount of material along the way. As the number of operations increases, the material loss rises and the probability of successful computation decays. Our implementation uses relatively benign physical manipulation, and avoids certain "lossy" operations.

5.8 Experimental Investigations

In this section we describe the results of an experimental implementation of the parallel filtering model. In particular, we concentrate on testing the efficiency of the implementation of the *remove* operation, which is central to our model. Although we have not yet advanced to the stage of fully implementing an entire algorithm, the results obtained are promising. However, it is important to note that the implementation of the *remove* operation is completely separate in conceptual terms to the actual nature of the model. The success or failure of any particular implementation does not detract in any way from the power of the model.

Experimental objectives

The primary objectives of the experiments detailed in this section are as follows:

1. To first ascertain optimal experimental conditions.
2. To test the implementation of the *remove* operation. We do this by performing a sequence of *removal* experiments, comprised of primer annealing, primer extension, and restriction.

Experimental overview

The primary objectives of the experiments detailed in this section are as follows:

1. To first ascertain optimal experimental conditions.
2. To test the implementation of the *remove* operation. We do this by performing a sequence of *removal* experiments, comprised of primer annealing, primer extension, and restriction.

3. To test the error-resistance of a *sequence* of removal operations that would
   occur during an actual algorithmic implementation.

Encoding colorings in DNA

We construct an initial library of (not necessarily legal) colorings in the following manner. For each vertex $v_i \in V$ we synthesize a *single* oligonucleotide (or
oligo) to represent each of $v_i =$ red, $v_i =$ green, and $v_i =$ blue. The structure
of these strands is depicted in Fig. 5.8.

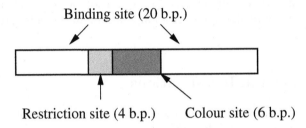

Fig. 5.8. Oligonucleotide structure

With reference to Fig. 5.8, each oligo (apart from those representing v_1 and
v_8) is composed of the following:

1. a unique 20-base binding site, within which is embedded a 4-base restriction site, $GATC$
2. a 6-base color identification site. The sequences chosen to represent "red",
   "green", and "blue" are $AAAAAA$, $GGGGGG$, and $CCCCCC$ respectively
3. a unique 20-base binding site

Oligos representing v_1 and v_8 are 34 bases long (8-base binding section, 6-
base color section and 20-base binding section). The initial sequences chosen
to represent each vertex coloring are listed in Table 5.1. Note that restriction
sites are underlined (e.g., $\underline{GATC}$) and color sequences are depicted in **bold**.
The melting temperature (T_m) of each strand is specified in the final column.

All sequences described here were originally designed by hand. This process is laborious and prone to error, and several researchers have described
subsequent attempts to automate the sequence design process. The sequences
were then checked with the Microgenie [127] package to check for common
subsequences and hairpin loops.

We now describe how the initial library is constructed from these oligos.
In order to reduce to a minimum the number of oligos to be synthesized, we
reject the *splinting* method described in [3, 98] in favor of an *overlapping*
approach.

Table 5.1. Sequences chosen to represent vertex/color combinations

Coloring	Sequence	T_m
$v_1 = red$	GCTCTGCT**AAAAAA**TCTTGATTTCACAGCATGGT	74.1
$v_1 = green$	GCTCTGCT**GGGGGG**TCTTGATTTCACAGCATGGT	83.1
$v_1 = blue$	GCTCTGCT**CCCCCC**TCTTGATTTCACAGCATGGT	82.5
$v_2 = red$	CGTCATAG<u>GATC</u>ACCATGCT**TTTTTT**TACCATGCTGTGAAATCAAGA	81.5
$v_2 = green$	CGTCATAG<u>GATC</u>ACCATGCT**CCCCCC**ACCATGCTGTGAAATCAAGA	88.4
$v_2 = blue$	CGTCATAG<u>GATC</u>ACCATGCT**GGGGGG**ACCATGCTGTGAAATCAAGA	88.4
$v_3 = red$	AGCATGGT<u>GATC</u>CTATGACG**AAAAAA**TGCTGCTAAGACGAAGAGTT	80.9
$v_3 = green$	AGCATGGT<u>GATC</u>CTATGACG**GGGGGGG**TGCTGCTAAGACGAAGAGTT	86.6
$v_3 = blue$	AGCATGGT<u>GATC</u>CTATGACG**CCCCCC**TGCTGCTAAGACGAAGAGTT	86.7
$v_4 = red$	GTAGGTGT<u>GATC</u>CAGTGGT**TTTTTTTT**AACTCTTCGTCTTAGCAGCA	79.2
$v_4 = green$	GTAGGTGT<u>GATC</u>CAGTGGTT**CCCCCC**AACTCTTCGTCTTAGCAGCA	86.0
$v_4 = blue$	GTAGGTGT<u>GATC</u>CAGTGGTT**GGGGGG**AACTCTTCGTCTTAGCAGCA	86.0
$v_5 = red$	AACCACTG<u>GATC</u>ACACCTAC**AAAAAA**GGTCTTCGGCGGCAATCTAC	83.7
$v_5 = green$	AACCACTG<u>GATC</u>ACACCTAC**GGGGGGGG**TCTTCGGCGGCAATCTAC	89.9
$v_5 = blue$	AACCACTG<u>GATC</u>ACACCTAC**CCCCCCC**GGTCTTCGGCGGCAATCTAC	89.9
$v_6 = red$	GTAGGTGT<u>GATC</u>CAGTGGT**TTTTTTTT**GTAGATTGCCGCCGAAGACC	83.8
$v_6 = green$	GTAGGTGT<u>GATC</u>CAGTGGTT**CCCCCC**GTAGATTGCCGCCGAAGACC	89.5
$v_6 = blue$	GTAGGTGT<u>GATC</u>CAGTGGTT**GGGGGGG**TAGATTGCCGCCGAAGACC	89.5
$v_7 = red$	AACCACTG<u>GATC</u>ACACCTAC**AAAAAA**CACTGACAAGACCTTTGCTT	80.8
$v_7 = green$	AACCACTG<u>GATC</u>ACACCTAC**GGGGGG**CACTGACAAGACCTTTGCTT	87.4
$v_7 = blue$	AACCACTG<u>GATC</u>ACACCTAC**CCCCCCCC**ACTGACAAGACCTTTGCTT	86.8
$v_8 = red$	GCGGAATTCCTCTGCTGATC**TTTTTTT**AAGCAAAGGTCTTGTCAGTG	81.9
$v_8 = green$	GCGGAATTCCTCTGCTGATC**CCCCCCC**AAGCAAAGGTCTTGTCAGTG	89.1
$v_8 = blue$	GCGGAATTCCTCTGCTGATC**GGGGGG**AAGCAAAGGTCTTGTCAGTG	89.1

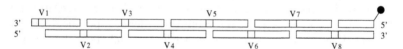

Fig. 5.9. Library construction

Sequences representing *odd*-numbered vertices run in the 3' → 5' direction. Sequences representing *even*-numbered vertices run in the 5' → 3' direction (see Fig. 5.9). We now describe the structure of the binding sections of strands representing even-numbered vertices. The "left-hand" binding section of strands representing v_n (where n is even) is the complement of the "right-hand" binding section of strands representing v_{n-1}. Similarly, the "right-hand" binding section of strands representing v_n is the "left-hand" binding section of strands representing v_{n+1}.

We also construct a single biotinylated oligo, corresponding to the complement of the "right-hand" binding section of strands representing v_8. This allows us to purify strands encoding colorings away from the splint strands. The use of hybridization extraction does not cause a problem at this stage, since the process is performed only once, rather than repeatedly during the main body of a computation.

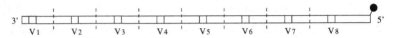

Fig. 5.10. Structure of strands in initial library

We then pour all oligos into solution. When all oligos have annealed we expect to obtain many double strands of the form depicted in Fig. 5.9. Vertex and color sections each occupy a distinct subsection of each strand. It is clear from the structure of these strands that the sequence encoding a particular vertex/color combination depends not only on the vertex in question, but on whether or not the vertex number is odd or even. For example, sections coloring v_1 "red" have the sequence $AAAAAA$, as expected. However, due to the overlapping nature of the strand construction technique, sections coloring v_2 have the sequence $TTTTTT$. This minor complication does not present a problem and, knowing the sequence assigned to each vertex/color combination, it is a trivial task to derive the sequences of the appropriate primer. The primer sequences are listed in Table 5.2.

Materials and methods

We now describe in detail the 32 experiments carried out during this particular phase of the project. A summary of these is given in Table 5.3. The results of these experiments are listed in the next section; here we describe only the materials and methods used. We represent the order of experimental execution in Fig. 5.11. Each process box is labelled with the numbers of the experiments carried out at that stage. Note that the only cycle in the flowchart occurs while attempting to remove strands containing a certain sequence. This is necessitated by the need for control and optimization experiments in order to establish optimal (or near-optimal) experimental conditions.

Experiment 1. Oligos and reagents

All the tile oligos were resuspended to 100 pmoles/μl in distilled water, then diluted to produce two mastermixes, the first containing all the oligos and the other containing only red oligos. The final concentration in each case was 2.5 pmoles each oligo/μl. The primer oligos were re-suspended to 100 pmoles/μl stocks and also diluted to 30 pmoles/μl PCR working stock. The biotinylated primer (v_8) was re-suspended to 200 pmoles/μl and stored in 20 μl single use aliquots. All oligos were stored at -20°C. A 5xPolymerase/Ligase buffer was made up (based on the requirements for second strand cDNA synthesis).

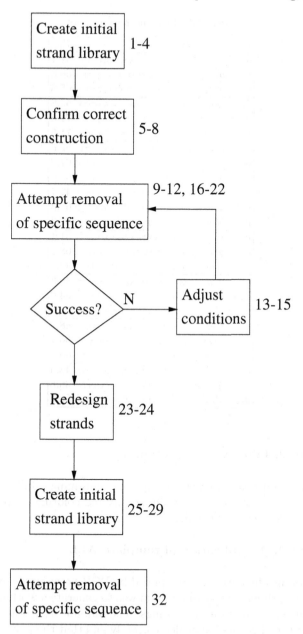

Fig. 5.11. Flowchart depicting experimental cycle

Table 5.2. Sequences of primers

Primer	Sequence	T_m
$v_1 = red$	GCTCTGCTAAAAAATCTT	51.1
$v_1 = green$	GCTCTGCTGGGGGGTCT	66.8
$v_1 = blue$	GCTCTGCTCCCCCCTCTT	65.8
$v_2 = red$	AGCATGGTAAAAAAAGCA	55.8
$v_2 = green$	AGCATGGTGGGGGGAGC	71.5
$v_2 = blue$	AGCATGGTCCCCCCAGCA	71.5
$v_3 = red$	CTATGACGAAAAAATGCT	52.6
$v_3 = green$	CTATGACGGGGGGGTGCT	67.2
$v_3 = blue$	CTATGACGCCCCCCTGCT	67.4
$v_4 = red$	GAAGAGTTAAAAAAAACC	47.7
$v_4 = green$	GAAGAGTTGGGGGGAACC	63.3
$v_4 = blue$	GAAGAGTTCCCCCCAACC	63.3
$v_5 = red$	ACACCTACAAAAAAGGTC	51.1
$v_5 = green$	ACACCTACGGGGGGGGTC	68.2
$v_5 = blue$	ACACCTACCCCCCCGGTC	49.3
$v_6 = red$	CAATCTACAAAAAAAACC	49.3
$v_6 = green$	CAATCTACGGGGGGAACC	64.0
$v_6 = blue$	CAATCTACCCCCCCAACC	63.9
$v_7 = red$	ACACCTACAAAAAACACT	48.7
$v_7 = green$	ACACCTACGGGGGGCACT	66.7
$v_7 = blue$	ACACCTACCCCCCCCACT	65.4
$v_8 = red$	CTTTGCTTAAAAAAGATC	48.9
$v_8 = green$	CTTTGCTTGGGGGGGAT	66.6
$v_8 = blue$	CTTTGCTTCCCCCCGAT	66.7

Experiment 2. Library construction

Two hybridization reactions were set up, one containing a full set of coloring oligos and the other a control containing only red coloring oligos. The products were labelled ALL and RED.

Experiment 3. Amplification of template ALL

Hybridization product ALL was amplified by PCR between v_8 and a mixture of the three v_1 primers. The primary aim was to generate a working stock containing a mixed population of colorings. The amplification was also used to confirm that the oligos had annealed correctly and that the polymerase/ligase step had repaired the gaps between oligos to produce double-stranded tile chains of the correct length. PCR template concentration and MgCl2 concentrations were titrated in order to optimize PCR conditions.

Table 5.3. Summary of experimental objectives

No.	Summary of experimental objective
1	Resuspension of oligos
2	Creation of initial library
3	Amplification of initial library
4	Purification of product of Experiment 3
5	Check to ensure correct hybridization
6	Test of specificity of Experiment 5
7–8	Cloning and sequencing to ensure correct library construction
9	Removal experiment 1
10	Amplification and purification of test library subset
11	Removal experiment 2
12	Removal experiment 3
13	Test of ability of Sau3A to digest dsDNA
14–15	Control experiments using MboI rather than Sau3A
16	Removal experiment 4
17	Removal experiment 5
18	Removal experiment 6
19	Removal experiment using Klenow (7)
20	Removal experiment 8
21	Removal experiment 9
22	Test of multiple removals
23–24	Oligo redesign
25	New initial library construction
26–29	New initial library amplification
30–31	PCR control experiments for new library
32	Removal experiment 10

Experiment 4. Purification of v_1–v_8 PCR product

The 10 best samples producing bands in Experiment 3 were pooled, and gel purified on 2% agarose. The $\sim$200 b.p. fragment was extracted using the Qiagen gel extraction kit. The PCR product was ligated into a T cloning vector and used to transform stored competent bacteria. Several thousand clones were produced. 12 were grown, miniprepped, and sequenced (Experiments 7 and 8).

Experiment 5. PCR between v_2 and v_8

The aim was to check that all three colorings of v_2 were present in the amplified chain produced in Experiment 3. This was done simply by using v_8 and either red, green, or blue specific v_2 primers in a standard detection PCR reaction.

Experiment 6. PCR from RED template (from Experiment 2)

This control experiment was set up to check the specificity of the PCR detection step (to make sure the color-specific primers did not cross-react, giving false positive results). The RED hybridization product from Experiment 2 was diluted 1/10 and color-specific detection PCRs set up for v_1, v_2, v_3, and v_4.

Experiments 7 and 8. Sequencing of clones from Experiment 4

Eight clones were sequenced using either universal or reverse primers.

Experiment 9. Exclusion experiment 1

This initial control experiment was designed to test the ability of the proposed Taq-based primer extension/Sau3A digestion method of excluding specific sequences from a mixed population of chains. Template prepared in Experiment 4 was bound to Dynabeads, denatured and then split into two. One half was treated with the intention of excluding all but one vertex coloring, the other was used as a control and taken through the procedure without adding any exclusion primers. Following Sau3A digestion, the surviving DNA was harvested from the beads by EcoR1 digestion. Detection PCR reactions were set up to detect specific vertex sequences within each of the samples. The untreated half was intended as a positive control for the detection PCR step.

Experiment 10. Amplification and purification of the RED template

A red-only template was prepared by PCR from the RED hybridization product (Experiment 2) using $v_1 = red$ and v_8 primers. The PCR product was gel purified and extracted using a Qiagen gel extraction kit.

Experiment 11. Exclusion experiment 2

This was a repeat of Experiment 9 with various modifications intended to increase the stringency of the exclusion step (amount of template reduced, primer concentration increased, annealing temperature reduced, number of cycles increased). The RED control template was also taken through the procedure as a PCR specificity control.

Experiment 12. Exclusion experiment 3

Conditions were modified further to favor exclusion. An additional control was included to test the ability of the Sau3A to destroy double-stranded sequences.

This sample was bound to the Dynabeads, washed but not denatured, and then taken through the procedure as double-stranded template, which should have been completely destroyed by the exclusion procedure.

Experiment 13. *Sau*3A control

This control was used to assess the ability of *Sau*3A to digest double-stranded DNA. Serial dilutions were made from template ALL. 1 μl of each dilution was digested using 10U *Sau*3A at 37°C for one hour, in a total volume of 10 μl. 1 μl of each digest, and undigested controls, were used as template in standard v_1–v_8 detection PCR reactions.

Experiments 14 and 15. Control experiments using Mbo1 to digest targeted sequences

Experiment 13 was repeated using *Mbo*1 (a *Sau*3A isoschizomer that has the same specificity for double stranded DNA).

Experiment 16. Exclusion experiment 4

Experiment 12 was repeated, substituting *Mbo*1 for *Sau*3A.

Experiment 17. Exclusion experiment 5

Fresh v_1–v_8 template was prepared and purified, then used in an exclusion experiment as in Experiment 12. Digestion was carried out overnight using 50U enzyme.

Experiment 18. Exclusion experiment 6

Template prepared in Experiment 17 was diluted 1/10 and 1 μl bound to Dynabeads, denatured, and washed. A v_2 exclusion was set up, using *Taq* in the primer extension reaction as usual. The beads were re-suspended every 10 cycles and a 3 μl aliquot of beads removed at roughly 20 cycle intervals, up to a maximum of 85 cycles. *Mbo*1 digestion of each aliquot was carried out overnight at 37°C in a rotating oven (to keep the beads in suspension) using 50U (a huge excess) of enzyme. v_2 detection PCRs were set up from each of the digested samples.

Experiment 19. Exclusion experiment using Klenow

In this experiment Klenow was substituted for *Taq* in the primer extension step. The idea was to overcome the problem of the beads settling out while cycling in the PCR block, which may have accounted for the inefficiency of the reaction. To do this the tubes were incubated in a 37°C rotating oven for three hours. Detection PCR was carried out as normal and the products run out alongside the samples from Experiment 18.

Experiment 20. Exclusion experiment 8

This was a Taq-based exclusion using modified cycling conditions. The ALL template from Experiment 17 was diluted 1/10 and bound to the Dynabeads, denatured, washed, and split into three. Three single exclusion reactions were set up, one each for the v_2 colors (in this way the three templates acted as PCR controls for each other).

Experiment 21. Exclusion experiment 9

Experiment 20 was repeated with modifications to primer extension conditions aiming to favor the annealing of red primers.

Experiment 22. Multiple tile exclusion

This was the first attempt to exclude more than one tile at a time. It was thought that expanding the experiment in this way may increase the efficiency of the exclusion reactions by reducing the overall level of template available at the detection PCR step.

Experiments 23 and 24. Oligo redesign

New oligos were designed to replace the v_3 oligos in the existing chain. By slotting in oligos with modified characteristics, it was hoped that the principle of the exclusion technique could be shown to work (even if it was not possible to execute a full algorithm). The basic design features of the new oligos were as follows:

- The regions of overlap with v_2 and v_4 were conserved so that the new oligos could be incorporated into the old chain.
- The color signatures were increased from six to ten nucleotides in length to increase stability.
- The three color sequences shared no base homology, i.e., they were different at each individual base.
- All the oligos were checked for runs of bases, homologies, hairpins, etc.
- Primer T_m were as close as possible to each other (so that primers could be used in combination under optimal conditions), and all around 60°C. This was achieved by maintaining a GC ratio of 50% for each primer.

The new sequences are described later in this section.

Experiment 25. New full-length chain preparation

Three new full-length chains were prepared containing either red, green, or blue new v_3 in a backbone of the original red tiles v_1, v_2, v_4, v_7, and v_8. By omitting the green and blue tiles in the backbone of the molecule we hoped to avoid any interactions between blue $CCCCCC$ and green $GGGGGG$ sequences, which could lead to the formation of hairpins in the full-length molecule. The construction method was exactly as in Experiment 2.

Experiments 26 to 29. New chain amplification

Each of the new chains were amplified and purified separately to produce both biotinylated and non-biotinylated products. The non-biotinylated products were cloned for sequencing (but unfortunately the sequencing reactions failed and there was not enough time to repeat them). The biotinylated products were used in the control and exclusion optimization experiments.

Experiments 30 and 31. PCR control experiments for the new tile chains

The three new tile chains containing either the red, green, or blue v_3 were assessed separately. Serial dilutions were made from each template (down to a dilution of 10:8). PCR reactions were then set up in order to determine the limit of detection for each template. For subsequent control and exclusion reactions these templates were used at a concentration approximately ten-fold above the limit of detection. It was hoped that by balancing the initial amount of template used in the experiments, any partial exclusion (as seen previously) would be sufficient to reduce the level of template below the limit of PCR detection, giving a negative result (i.e., showing that exclusion had been successful). PCR conditions were optimized to ensure that there was no cross reaction between the new colored sequences, and $Mbo1$ digestion times were assessed to ensure complete digestion of double-stranded DNA at these concentrations.

Experiment 32. Final exclusion experiment

Red, green, and blue templates were mixed in roughly equal proportions (based on their PCR detection limit in Experiment 30). The mixed template was bound to the Dynabeads and prepared in bulk before splitting into four tubes. Red, green, and blue v_3 exclusions were set up in separate reactions, alongside a no-exclusion control. Following $Mbob$ digestion, detection PCR reactions were set up to determine the relative levels of each v_3 in each of the exclusion samples and control.

Results obtained

Experiment 3. The major product of ~200 b.p. was detectable at all template concentrations, though very faint at 1/1000 dilution. MgCl2 concentration had very little effect on PCR efficiency. Optimal conditions appeared to be at 2mM MgCl2 using the template diluted to 1/10. In all cases the product appeared slightly smaller than expected, though it was not clear if this was a gel artifact or a problem with the library oligo assembly.

Experiment 5. All three vertices were found to be represented in the chain population (assuming no cross reaction had occurred)

Experiment 6. The PCR products were faint but in each case appeared to be color-specific. There was also a stepwise reduction in the size of the PCR product from v_1 through v_4, showing that the PCR reactions were vertex-specific and that the majority of colorings had assembled in the correct order.

Experiments 7 and 8. It was clear that there was fairly high sequence variability among the clones. They showed a number of vertex assembly patterns and chain lengths. Some could be explained by PCR mispriming during the chain amplification step, yielding products with a v_1 sequence at both ends (these products would not cause problems in the exclusion experiments since they were not biotinylated, would not bind to the Dynabeads, and would be removed during the washing step). One feature common to all the clones was the absence of sequences representing v_5 and v_6. Looking at the oligo sequences it was clear that the problem was due to identical overlapping regions between v_4 and v_5, and v_6 and v_7, making two chains possible, the shorter of which seemed to form predominantly. In this small selection of clones it looked like the vertex colorings were represented equally among the chains.

Experiment 9. The detection reactions showed that the PCR products from each sample were of equal intensity for each vertex tested, showing that exclusion under these conditions had failed.

Experiment 11. The PCR results showed that the detection step was specific, but that the exclusion steps had not reduced the amount of targeted sequence.

Experiment 12. Detection PCR results showed that the PCR was specific (RED control), but that the exclusion steps had not worked, and also that Sau3A had failed to destroy the double-stranded control. This implied that the exclusion experiments were failing due to incomplete Sau3A digestion of the marked sequence.

Experiment 13. Comparisons of the two sets of samples showed that digestion with Sau3A made no difference to the detection limit of the PCR reactions. As a further control, the products of the above reactions were gel purified and split in two. One half was digested with Sau3A and visualized alongside the other (undigested) control on 2% agarose. The undigested DNA ran as a distinct band, whereas the digested half appeared as a high molecular weight smear.

Experiments 14 and 15. An overnight 37°C digestion using 20U of enzyme was found to completely destroy the template (i.e., to reduce the level of template below the limit of PCR detection).

Experiment 16. Nothing could be concluded from this set of reactions since the positive PCR controls failed.

Experiment 17. The positive and negative detection PCR controls worked, but all other reactions failed. The problem seemed to be due to inefficient harvesting of the excluded template from the Dynabeads prior to detection PCR. To get round this problem an unbiotinylated v_8 primer was ordered. Using this primer, detection PCRs could be set up directly from templates bound to the Dynabeads.

Experiment 18. The results of this experiment gave the first evidence that the exclusion method could work. The intensity of the specific PCR product band decreased with increased number of exclusion cycles, although the exclusion never reached completion. The template was still detectable after 85 cycles of primer extension.

Experiment 19. The use of Klenow produced the same effect as ~30 to 40 cycles of Taq-based exclusion (i.e., exclusion was not complete), showing that Klenow offered no advantage over Taq. Detection PCR showed specific exclusion of $v_2 = green$ and $v_2 = blue$ sequences, but not $v_2 = red$. The gel is depicted in Fig. 5.12. A summarized interpretation of this gel is presented in Table 5.4.

Lanes 1 to 3 show the result of the removal of strands encoding $v_2 = red$. Lane 1, corresponding to $v_2 = red$ *should* be empty, but a faint band is visible. Lanes 2 and 3, corresponding to $v_2 = green$ and $v_2 = blue$ primers respectively contain normal length product, showing that strands *not* containing the sequence $v_2 = red$ were not removed.

We believe that the incomplete removal of $v_2 = red$ strands is due to the sequence chosen to represent red ($AAAAAA$). Because adenine only forms two hydrogen bonds with thymine, the optimum annealing temperature between strands and red primers is lower than that for green ($CCCCCC$) and blue ($GGGGGG$) primers. We believe a simple modification to the encoding sequence (described later) will solve this problem.

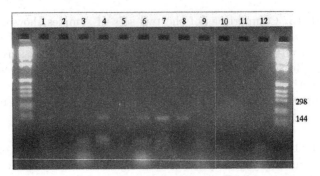

Fig. 5.12. Visualization of gel resulting from Experiment 20

Lanes 4 to 6 show the result of the removal of strands encoding $v_2 = green$. Lane 5, corresponding to the $v_2 = green$ primer, is empty, showing that no strands containing that sequence were present. Lanes 4 and 6, corresponding to $v_2 = red$ and $v_2 = blue$ primers respectively contain normal length product, showing that strands *not* containing the sequence $v_2 = green$ were not removed.

Lanes 7 to 9 show the result of the removal of strands encoding $v_2 = blue$. Lane 9, corresponding to the $v_2 = blue$ primer is empty, showing that no strands containing that sequence were present. Lanes 7 and 8, corresponding to $v_2 = red$ and $v_2 = green$ primers respectively contain normal length product, showing that strands *not* containing the sequence $v_2 = blue$ were not removed.

The streaks visible at 74 and 18 b.p. are due to the presence of primer dimers and free primers respectively.

Table 5.4. Interpretation of Fig. 5.12

Lane	Excluded	PCR Primer	Result
1		$v_2 = $ red	+
2	$v_2 = $ red	$v_2 = $ green	+
3		$v_2 = $ blue	+
4		$v_2 = $ red	+
5	$v_2 = $ green	$v_2 = $ green	-
6		$v_2 = $ blue	+
7		$v_2 = $ red	+
8	$v_2 = $ blue	$v_2 = $ green	+
9		$v_2 = $ blue	-
10	None	$v_2 = $ red	-
11	(PCR -ve	$v_2 = $ green	-
12	control)	$v_2 = $ blue	-

Experiment 21. Evidence of exclusion was seen again, but in all cases it was incomplete.

Experiment 22. There was evidence of specific exclusion (the intensity of targeted sequences reduced) but the process was incomplete. There seemed to be a basic problem with the method in that it used the enzymatic removal process to target and destroy specific sequences, followed by an incredibly sensitive technique to detect them.

Experiment 32. The PCR failed to produce any product from any of the samples, including the positive control. This was probably due to loss of the template during the washing steps, reducing its concentration below the limit of detection.

Discussion

It is perhaps useful at this point to note that this implementation, first proposed in [12], established for the first time the "destructive" DNA-based algorithmic paradigm, which has been subsequently used in several ground-breaking papers [58, 99].

In [141] Seeman et al. describe the potential pitfalls that may confront experimentalists working on DNA computation. In this section we describe in a similar fashion the lessons to be drawn from the experimental investigations just described. We hope that other experimentalists in the field may be made aware of various subtle aspects of the implementation of models of DNA computation. We have found that the requirements of DNA-based algorithmic experiments are often more strict than those of "traditional" investigations in molecular biology. For example, it is rare that molecular biologists are required to sequence a heterogeneous population of DNA strands; yet, for any nontrivial problem, this task is inevitably required as the final step of the implementation of a DNA-based algorithm. We hope that these (often non-obvious) impediments to efficient and error-resistant implementation of models of DNA computation will be made apparent in the following sections.

Ensure appropriate control and optimization experiments are performed

We quickly found that a major component of the work was comprised of finding optimal experimental conditions. Factors to be taken into account included strand concentration, salt concentration, restriction enzyme concentration, annealing temperature, and number of cycles. Due to the unusual nature of the experiments, we found that the system was far more sensitive to experimental conditions than is normally the case.

We also carried out extensive control experiments to ensure the specificity of the PCR detection step (i.e., to ensure that strands were not removed without this being done explicitly). Also, control experiments indicated the inefficiency of the $Sau3A$ restriction enzyme that was originally used, though an isoschizomer, $MboI$, worked well.

Ensure that the initial library is constructed cleanly before proceeding

A fundamental prerequisite for correct algorithmic implementation is that the initial library of strands be constructed as expected. This is especially important for algorithms within filtering models, since we must be absolutely sure that every possible solution to the given problem is represented as a strand. While describing their attempt to recreate Adleman's experiment, Kaplan et al. [86] acknowledge the difficulty of obtaining clean generation of the initial library.

There are several potential problems inherent to the construction of an initial library by the annealing and ligation of many small strands. Incomplete or irregular ligation can result in shorter than expected strands. We checked for this, and observed that the majority of the product was of the expected length. In addition to checking the length of the product, we rigorously ensured that there is sufficient variability within the initial library by cloning a sample into $E.coli$ and sequencing their DNA.

Correct strand/primer design is vital

In [3] Adleman originally suggested using random sequences to represent vertices within the given graph. He explained this choice by stating that it was unlikely that sequences chosen to represent different vertices would share long common subsequences, and that undesirable features such as hairpin loops would be unlikely to occur. The selection of random sequences was also supported by Lipton in [98].

Since the publication of [3] and [98], the use of random sequences has been called into question [23, 51, 52, 108]. It is clear that for any nontrivial problem, careful thought must go into the design of sequences to represent potential solutions if we are to avoid the problems described above.

One major problem we encountered was due to the sequences chosen to represent target sequences. We made a completely arbitrary decision to differentiate S_1 by the sequence $AAAAAA$, S_2 by $CCCCCC$, and S_3 by $GGGGGG$. In retrospect, it is clear that this was a bad choice for two main reasons. The first concerns the S_2 and S_3 primers. It is clear that, in solution, these primers are complementary, and are just as likely to anneal to one another as they are to the target sequences. Obviously, this will greatly reduce the efficiency of the removal operation. The second problem concerns the melting temperatures of the primers. Because the melting temperatures of the S_1 primers was far lower than that of the S_2 and S_3 primers, we observed incomplete removal of S_1 sequences. As a result of these problems, we redesigned the strands, the modifications being detailed in [6].

PCR can introduce problems

It is unrealistic to assume that our enzymatic method removes 100% of the targeted strands. We must therefore be prepared to accept that a small proportion of target strands will be left in solution. Normally, this residue would be undetectable, but the repeated use of PCR can quickly amplify this trace amount, causing failure of the algorithm being implemented. The experiment confirmed that our removal method worked, but the use of PCR as a detection method was far too sensitive for our purposes. Kaplan et al. [87] confirm our belief that PCR is a major source of errors.

Biotinylated strands can introduce problems

In our experiments we used a biotinylated primer to purify away the "splint" strands used to construct the initial library. Quite apart from the problems with biotinylation described earlier, it became clear from our investigations that this can cause other significant difficulties. We found that the attached beads "settled out" in solution, dragging the strands to the bottom of the heating block and affecting the efficiency of the process. We overcame this problem by incubating the tubes in a rotating oven.

Restriction enzymes are often not as effective as they are claimed to be

Although various claims are made for the efficiency of restriction enzymes, in reality they have a nonzero error rate associated with them. We found that Sau3A was ineffective at cleaving double-stranded DNA, but that MboI worked perfectly well. This may have been due to the fact that Sau3A is inefficient at cleaving $de\ novo$ synthesized DNA.

5.9 Other Laboratory Implementations

In this section we describe several successful laboratory implementations of molecular-based solutions to NP-complete problems. The objective is not to give an exhaustive description of each experiment, but to give a high-level treatment of the general methodology, so that the reader may approach with confidence the fuller description in the literature.

5.9.1 Chess Games

In [58], Faulhammer et al. describe a solution to a variant of the satisfiability problem that uses RNA rather than DNA as the computational substrate. They consider a variant of SAT, the so-called "Knight problem", which seeks configurations of knights on an $n \times n$ chess board, such that no knight is attacking any other. Examples of legal and illegal configurations are depicted in Fig. 5.13.

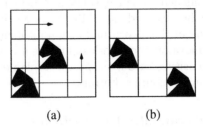

(a)   (b)

Fig. 5.13. (a) Legal configuration. (b) Illegal configuration - both knights can attack one another

In keeping with our earlier observations, the authors prefer a "mark and destroy" strategy [12] rather than the repeated use of hybridization extraction to remove illegal solutions. However, the use of an RNA library and ribonuclease (RNase) H digestion gives greater flexibility, as one is not constrained by the set of restriction enzymes available. In this way, the RNase H acts as a "universal restriction enzyme", allowing selective marking of virtually any RNA strands for parallel destruction by digestion.

The particular instance solved in [58] used a 3×3 board, with the variables $a - i$ representing the squares (Fig. 5.14). If a variable is set to 1 then a knight is present at that variable's square, and 0 represents the absence of a knight.

a	b	c
d	e	f
g	h	i

Fig. 5.14. Labelling of test board

The 3×3 knight problem may therefore be represented as the following instance of SAT:

$$((\neg h \wedge \neg f) \vee \neg a) \wedge ((\neg g \wedge \neg i) \vee \neg b) \wedge ((\neg d \wedge \neg h) \vee \neg c) \wedge ((\neg c \wedge \neg i) \vee \neg d) \wedge$$
$$((\neg a \wedge \neg g) \vee \neg f) \wedge ((\neg b \vee \neg f) \vee \neg g) \wedge ((\neg a \wedge \neg c) \vee \neg h) \wedge ((\neg d \wedge \neg b) \vee \neg i)$$

which, in this case, simplifies to

$$((\neg h \wedge \neg f) \wedge \neg a) \wedge ((\neg g \wedge \neg i) \vee \neg b) \wedge ((\neg d \wedge \neg h) \vee \neg c) \wedge ((\neg c \wedge \neg i) \vee \neg d) \wedge$$
$$((\neg a \wedge \neg g) \vee \neg f).$$

This simplification greatly reduces the number of laboratory steps required. The experiment proceeds by using a series of RNase H digestions of "illegal" board representations, along the lines of the parallel filtering model [12].

Board representations are encoded as follows: the experiment starts with all strings of the form $x_1, \ldots, x_n$, where each variable x_i takes the value 1 or 0; then, the following operations may be performed on the population of strings:

- Cut all strings containing any pattern of specified variables $p_i, \ldots, p_k$
- Separate the "test tube" into several collections of strings (molecules) by length
- Equally divide (i.e., split) the contents of a tube into two tubes
- Pour (mix) two test tubes together
- Sample a random string from the test tube

The first stage of the algorithm is the construction of the initial library of strands. Each strand sequence follows the template depicted in Fig. 5.15.

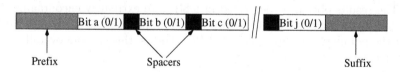

Fig. 5.15. Template for RNA strands

The prefix and suffix regions are included to facilitate PCR. Each variable is represented by one of two unique sequences of length 15 nucleotides, one representing the fact that the variable is set to 1, and the other the fact that it is set to 0. Variable regions are separated by short (5 nucleotide) spacer regions. In order to avoid having to individually generate each individual sequence, a "mix and split" strategy (described in more detail in [58]) is used. The RNA version of the library is then generated by *in vitro* transcription. The algorithm proceeds as follows:

1. For each square, sequentially, split the RNA library into two tubes, labelled 1 and 2. After digestions have taken place, tube 1 will contain

strands that contain a knight at that square, and tube 2 will contain strands that do *not* have knights at that square

2. In tube 1, digest with RNase H strands that have no knight at position **a**, as well as strands that describe a knight at attacking positions **h** and **f**. This implements the logical statement $((\neg h \wedge \neg f) \vee \neg a)$

3. In tube 2, digest strands that have a knight present at position **a**

4. Remove the DNA oligos used to perform the above digestions

5. Go to step 1, repeating with square **b**

Steps 1 through 4 implement the following: "There may or may not be a knight in square **a**: if there *is*, then it is attacking squares **h** and **f**, so disallow this." The algorithm only needs to be performed for squares **a, b, c, d,** and **f**, as square **e**, by the rules of chess, cannot threaten or be threatened on a board this size, and any illegal interactions that squares **g, h,** and **i** may have are with **a, b, c, d,** and **f**, and have already been dealt with. At the conclusion of this stage, any remaining full-length strands are recovered, as they should encode legal boards.

The "mark and destroy" digestion operation is implemented as follows. If we wish to retain (i.e., select) strands encoding variable a to have value 1, DNA oligonucleotides corresponding to the complement of the $a = 0$ sequence are added to the tube, and anneal to all strands encoding $a = 0$. RNase H is then added to the solution. Ribonuclease H (RNase H) is an endoribonuclease which specifically hydrolyzes the phosphodiester bonds of RNA hybridized to DNA. RNase H does not digest single or double-stranded DNA, so his operation therefore leaves intact only those strands encoding $a = 1$, in a fashion similar to the removal operation of the parallel filtering model [12].

The results obtained (described in [58]) were extremely encouraging: out of 43 output strands sampled, only one contained an illegal board. Given that the population sampled encoded 127 knights, this gave an overall knight placement success rate of 97.7%.

5.9.2 Computing on Surfaces

Another experiment that makes use of the "mark and destroy" paradigm [12] is described in [99]. The key difference between this and previous experiments is that the DNA strands used are tethered to a support rather than being allowed to float freely in solution. The authors argue that this approach greatly simplifies the automation of the (potentially very many) repetitive chemical processes required during the performance of an experiment.

The authors report a DNA-based solution to a small instance of the SAT problem (described in Chap. 2). The specific problem solved is

$$(w \vee x \vee y) \wedge (w \vee \neg y \vee z) \wedge (\neg x \vee y) \wedge (\neg w \vee \neg y).$$

16 unique DNA strands were synthesized, each one corresponding to one of
the $2^4 = 16$ combinations of variable values. The actual encodings are given
in Table 5.5 (taken from [99]).

Table 5.5. Strands used to represent SAT variable values

Strand	Sequence	$wxyz$
S_0	CAACCCAA	0000
S_1	TCTCAGAG	0001
S_2	GAAGGCAT	0010
S_3	AGGAATGC	0011
S_4	ATCGAGCT	0100
S_5	TTGGACCA	0101
S_6	ACCATTGG	0110
S_7	GTTGGGTT	0111
S_8	CCAAGTTG	1000
S_9	CAGTTGAC	1001
$S_1 0$	TGGTTTGG	1010
$S_1 1$	GATCCGAT	1011
$S_1 2$	ATATCGCG	1100
$S_1 3$	GGTTCAAC	1101
$S_1 4$	AACCTGGT	1110
$S_1 5$	ACTGGTCA	1111

Each of the 16 sets of strands was then affixed to a specific region of a gold
coated surface, so that each solution to the SAT problem was represented as
an individual cluster of strands.

The algorithm then proceeds as follows. For each clause of the problem,
a cycle of "mark", "destroy", and "unmark" operations is carried out. The
goal of each cycle is to destroy the strands that do *not* satisfy the appropriate
clause. Thus, in the first cycle, the objective is to destroy strands that do
not satisfy the clause $(w \vee x \vee y)$. Destruction is achieved by "protecting",
or *marking* strands that *do* satisfy the clause by annealing to them their
complementary strands. *E.coli* exonuclease I is then used to digest unmarked
strands (i.e., any single-stranded DNA).

By inspection of Table 5.5, we see that this applies to only two strands, S_0
(0000) and S_1 (0001). Thus, in cycle 1 the complements of the 14 other strands

$(w = 1(S_8, S_9, S_1 0, S_1 1, S_1 2, S_1 3, S_1 4, S_1 5);$
$x = 1(S_4, S_5, S_6, S_7, S_1 2, S_1 3, S_1 4, S_1 5);$
$y = 1(S_2, S_3, S_6, S_7, S_1 0, S_1 1, S_1 4, S_1 5))$

were combined and hybridized to the surface before the exonuclease I was
added. The surface was then regenerated (the unmark operation) to return
the remaining surface-bound oligos (S_2-$S_1 5$) to single-stranded form. This

process was repeated three more times for the remaining three clauses, leaving a surface containing only strands encoding a legal solution to the SAT problem.

The remaining molecules were amplified using PCR and then hybridized to an addressed array. The results of fluorescence imaging clearly showed four spots of relatively high intensity, corresponding to the four regions occupied by legal solutions to the problem (S_3, S_7, S_8, and S_9).

Although these results are encouraging as a first move toward more error-resistant DNA computing, as the authors themselves acknowledge, there remain serious concerns about the scalability of this approach (1,536 individual oligos would be required for a 36-bit, as opposed to 4-bit, implementation).

5.9.3 Gel-Based Computing

A much larger (20 variable) instance of 3-SAT was successfully solved by Adleman's group in an experiment described in [33]. This is, to date, the largest problem instance successfully solved by a DNA-based computer; indeed, as the authors state, "this computational problem may yet be the largest yet solved by nonelectronic means" [33].

The architecture underlying the experiment is related to the Sticker Model (see Chap. 3) described by Roweis et al. [133]. The difference here is that only separation steps are used – the application of stickers is not used. Separations are achieved by using oligo probes immobilized in polyacrylamide gel-filled glass modules, and strands are pulled through them by electrophoresis. Strands are removed (i.e., retained in the module) by virtue of their hybridizing to the immobilized probes, with other strands free to pass through the module and be subject to further processing. Captured strands may be released and transported (again via electrophoresis) to other modules for further processing.

The potential benefits of such an approach are clear; the use of electrophoresis minimizes the number of laboratory operations performed on strands, which, in turn, increases the chance of success of an experiment. Since strands are not deliberately damaged in any way, they, together with the glass modules, are potentially reusable for multiple computations. Finally, the whole process is potentially automatable, which may take us one step further towards a fully integrated DNA-based computer that requires minimal human intervention.

The problem solves was a 20-variable, 24-clause 3-SAT formula Φ, with a unique satisfying truth assignment. These are

$\Phi = (\neg x_{13} \lor x_{16} \lor x_{18}) \land (x_5 \lor x_{12} \lor \neg x_9) \land (\neg x_{13} \lor \neg x_2 \lor x_{20}) \land (x_{12} \lor x_9 \lor \neg x_5) \land (x_{19} \lor \neg x_4 \lor x_6) \land (x_9 \lor x_{12} \lor \neg x_5) \land (\neg x_1 \lor x_4 \lor \neg x_{11}) \land (x_{13} \lor \neg x_2 \lor \neg x_{19}) \land (x_5 \lor x_{17} \lor x_9) \land (x_{15} \lor x_9 \lor \neg x_{17}) \land (\neg x_5 \lor \neg x_9 \lor \neg x_{12}) \land (x_6 \lor x_{11} \lor x_4) \land (\neg x_{15} \lor \neg x_{17} \lor x_7) \land (\neg x_6 \lor x_{19} \lor x_{13}) \land (\neg x_{12} \lor \neg x_9 \lor x_5) \land (x_{12} \lor x_1 \lor x_{14}) \land (x_{20} \lor x_3 \lor x_2) \land (x_{10} \lor \neg x_7 \lor \neg x_8) \land (\neg x_5 \lor x_9 \lor \neg x_{12}) \land (x_{18} \lor \neg x_{20} \lor x_3) \land (\neg x_{10} \lor \neg x_{18} \lor \neg x_{16}) \land (x_1 \lor \neg x_{11} \lor \neg x_{14}) \land (x_8 \lor \neg x_7 \lor \neg x_{15}) \land (\neg x_8 \lor x_{16} \lor \neg x_{10})$

with a unique satisfying assignment of

$x_1 = F, x_2 = T, x_3 = F, x_4 = F, x_5 = F, x_6 = F, x_7 = T, x_8 = T, x_9 = F, x_{10} = T, x_{11} = T, x_{12} = T, x_{13} = F, x_{14} = F, x_{15} = T, x_{16} = T, x_{17} = T, x_{18} = F, x_{19} = F, x_{20} = F.$

As there are 20 variables, there are $2^{20} = 1,048,576$ possible truth assignments. To represent all possible assignments, two distinct 15 base *value sequences* were assigned to each variable $x_k (k = 1, \ldots, 20)$, one representing true (T), X_k^T, and one representing false (F), X_k^F. A mix and split generation technique similar to that of Faulhammer et al. [58] was used to generate a 300-base *library sequence* for each of the unique truth assignments. Each library sequence was made up of 20 value sequences joined together, representing the 20 different variables. These library sequences were then amplified with PCR.

The computation proceeds as follows: for each clause, a glass clause module is constructed which is filled with gel and contains covalently bound probes designed to capture only those library strands that *do satisfy* that clause; strands that do not satisfy the clause are discarded.

In the first clause module $(\neg x_{13} \lor x_{16} \lor x_{18})$ strands encoding X_3^F, X_{16}^F, and X_{18}^T are retained, while strands encoding X_{13}^T, X_{16}^F, and X_{18}^F are discarded. Retained strands are then used as input to the next clause module, for each of the remaining clauses. The final (24th) clause module should contain only those strands that have been retained in all 24 clause modules and hence encode truth assignments satisfying Φ.

The experimental results confirmed that a unique satisfying truth assignment for Φ was indeed found using this method. Impressive though it is, the authors still regard with scepticism claims made for the potential superiority of DNA-based computers over their traditional silicon counterparts. However, "they enlighten us about alternatives to electronic computers and studying them may ultimately lead us to the true 'computer of the future'" [33].

5.9.4 Maximal Clique Computation

The problem of finding a Maximal Clique (see Chap. 2) using DNA is addressed by Ouyang et al. in [115]. We recall that a clique is a fully connected subgraph of a given graph. The maximal clique problem asks: given a graph, how many vertices are there in the largest clique? Finding the size of the largest clique is an NP-complete problem.

The algorithm proceeds as follows: for a graph G with n vertices, all possible vertex subsets (subgraphs) are represented by an n-bit binary string $b_{n-1}, b_{n-2}, \ldots, b_0$. For example, given the six-vertex graph used in [115] and depicted in Fig. 5.16a, the string 111000 corresponds to the subgraph depicted in Fig. 5.16b, containing v_5, v_4, and v_3.

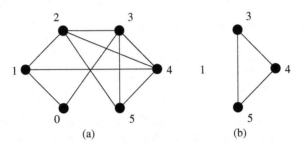

(a)   (b)

Fig. 5.16. (a) Ouyang graph. (b) Example subgraph

Clearly, the largest clique in this graph contains v_5, v_4, v_3, and v_2, represented by the string 111100.

The next stage is to find pairs of vertices that are not connected by an edge (and, therefore, by definition, cannot appear together in a clique). We begin by taking the *complement* of G, $\overline{G}$, which contains the same vertex set as G, but which only contains an edge $\{v, w\}$ if $\{v, w\}$ is *not* present in the edge set of G. The complement of the graph depicted in Fig. 5.16 is shown in Fig. 5.17.

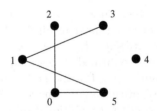

Fig. 5.17. Complement of Ouyang graph

If two vertices in $\overline{G}$ are connected by an edge then their corresponding bits cannot both be set to 1 in any given string. For the given problem, we must therefore remove strings encoding ***1*1 (v_2 and v_0), 1****1 (v_5 and v_0), 1***1* (v_5 and v_1) and **1*1* (v_3 and v_1), where * means either 1 or 0. All other strings encode a (not necessarily maximal) clique.

We must then sort the remaining strings to find the largest clique. This is simply a case of finding the string containing the largest number of 1s, as each 1 corresponds to a vertex in the clique. The string containing the largest number of 1s encodes the largest clique in the graph.

The DNA implementation of the algorithm goes as follows. The first task is to construct a set of DNA strands to represent all possible subgraphs. There are two strands per bit, to represent either 1 or 0. Each strand associated with a bit i is represented by two sections, its position P_i and its value V_i. All P_i sections are of length 20 bases. If $V_i = 1$ then the sequence representing V_i is a restriction site unique to that strand. If $V_i = 0$ then the sequence representing V_i is a 10 base "filler" sequence. Therefore, the longest possible sequence is 200 bases, corresponding to the string 000000, and the shortest sequence is 140 bases, corresponding to 111111. Parallel overlap assembly [85] was then used to construct the initial library.

The computation then proceeds by digesting strands in the library, guided by the complementary graph $\overline{G}$. To remove strands encoding a connection i, j in $\overline{G}$, the current tube is divided into two, t_0 and t_1. In t_0 we cut strings encoding $V_i = 1$ by adding the restriction enzyme associated with V_i. In t_1 we cut strings encoding $V_j = 1$ by adding the restriction enzyme associated with V_j. For example, to remove strands encoding the connection between V_0 and V_2, we cut strings containing $V_0 = 1$ in t_0 with the enzyme Afl II, and we cut strings containing $V_2 = 1$ in t_1 with Spe I. The two tubes are then combined into a new working tube, and the next edge in $\overline{G}$ is dealt with.

In order to read the size of the largest clique, the final tube was simply run on a gel. The authors performed this operation, and found the shortest band to be 160 bp, corresponding to a 4-vertex clique. This DNA was then sequenced and found to represent the correct solution, 111100.

Although this is another nice illustration of a DNA-based computation, the authors acknowledge the lack of scalability of their approach. One major factor is the requirement that each vertex be associated with an individual restriction enzyme. This, of course, limits the number of vertices that can be handled by the number of restriction enzymes available. However, a more fundamental issue is the exponential growth in the problem size (and thus the initial library), as has already been noted.

5.9.5 Other Notable Results

In this section we briefly introduce other notable experimental results, and bring attention to "late-breaking" results.

One of the first successful experiments reported after Adleman's result is due to Guarnieri et al. [71], in which they describe a DNA-based algorithm for binary addition. The method uses single-stranded DNA reactions to add together two nonnegative binary numbers. This application, as the authors note, is very different from previous proposals, which use DNA as the substrate for a massively parallel random search.

Adding binary numbers requires keeping track of the position of each digit and of any "carries" that arise from adding 1 to 1 (remembering that 1 +

$1 = 0$ plus carry 1 in binary). The DNA sequences used represent not only binary strings but also allow for carries and the extension of DNA strands to represent answers. Guarnieri et al. use sequences that encode a digit in a given position and its significance, or position from the right. For example, the first digit in the first position is represented by two DNA strands, each consisting of a short sequence representing a "position transfer operator", a short sequence representing the digit's value, and a short sequence representing a "position operator."

DNA representations of all possible two bit binary integers are constructed, which can then be added in pairs. Adding a pair involves adding appropriate complementary strands, which then link up and provide the basis for strand extension to make new, longer strands. This is termed a "horizontal chain reaction", where input sequences serve as templates for constructing an extended result strand. The final strand serves as a record of successive operations, which is then read out to yield the answer digits in the correct order.

The results obtained confirmed the correct addition of $0 + 0$, $0 + 1$, $1 + 0$, and $1 + 1$, each calculation taking between 1 and 2 days of bench work. Although limited in scope, this experiment was (at the time) one of the few experimental implementations to support theoretical results.

The tendency of DNA molecules to self-anneal was exploited by Sakamoto et al. in [135] for the purposes of solving a small instance of SAT. The authors encode the given formula in "literal strings" which are conjunctions of the literals selected from each SAT clause (one literal per clause). A formula is satisfiable if there exists a literal string that does not contain any variable together with its negation. If each variable is encoded as a DNA subsequence that is the Watson-Crick complement of its negation then any strands containing a variable and its negation self-anneal to form "hairpin" structures. These can be distinguished from non-hairpin structure-forming strands, and removed. The benefit of this approach is that it does not require physical manipulation of the DNA, only temperature cycling. The drawback is that it requires 3^m literal strings for m clauses, thus invoking once again the scalability argument.

Algorithmic self-assembly (as described in Chap. 3) has been demonstrated in the laboratory by Mao et al. [103]. This builds on work done on the self-assembly of periodic two-dimensional arrays (or "sheets") of DNA tiles connected by "sticky" pads [161, 163]. The authors of [103] report a one-dimensional algorithmic self-assembly of DNA triple-crossover molecules (tiles) to execute four steps of a logical XOR operation on a string of binary bits.

Triple-crossover molecules contain four strands that self-assemble through Watson-Crick complementarity to produce three double helices in roughly a planar formation. Each double helix is connected to adjacent double helices at points where their strands cross over between them. The ends of the core helix are closed by hairpin loops, but the other helices may end in sticky ends which direct the assembly of the macrostructure. The tiles then self-assemble to perform a computation. The authors of [103] report successful XOR computations on pairs of bits, but note that the scalability of the approach relies on proper hairpin formation in very long single-stranded molecules, which cannot be assumed.

We now briefly describe some "late-breaking" results. The construction of molecular automata (see Chap. 3) was demonstrated by Benenson et al. in [27]. This experiment builds on the authors' earlier work [28] on the construction of biomolecular machines. In [27], the authors describe the construction of a molecular automaton that uses the process of DNA backbone hydrolysis and strand hybridization, fuelled by the potential free energy stored in the DNA itself.

Related work, due to Stojanovic and Stefanovic [147], describes a molecular automaton that plays the game of tic-tac-toe (or noughts and crosses) against a human opponent. The automaton is a Boolean network of deoxribozymes incorporating 23 molecular-scale logic gates and one constitutively active deozyribozyme arrayed in a 3×3 well formation (to represent the game board). The human player signals a move by adding an input oligo, and the automaton's move is signalled by fluorescence in a particular well. This cycle continues until there is either a draw or a victory for the automaton, as it plays a perfect strategy and cannot be defeated.

5.10 Summary

In this chapter we have described in depth the experimental realization of some of the abstract models of DNA computation described in Chap. 2. We described Adleman's seminal experiment, as well as a potential implementation of the parallel filtering model, which laid the foundations for important later work on destructive algorithms. We also described some key contributions to the laboratory implementation of computations, and highlighted some late-breaking results.

5.11 Bibliographical Notes

The use of molecules other than DNA (for example, proteins and chemical systems) is reviewed and discussed in [144]. Chen and Wood [44] review early work on implementations of biomolecular computatons, and suggest potentially useful lines of enquiry. The recent proceedings of the International Workshop on DNA Based Computers [43, 73] contain many articles on laboratory implementations, including notable papers on whiplash PCR [105] and DNA-based memory [42].

6

Cellular Computing

> *"The punched tape running along the inner seam of the double helix is much more than a repository of enzyme stencils. It packs itself with regulators, suppressors, promoters, case-statements, if-thens."*
> *Richard Powers, The Gold Bug Variations* [122]

Complex natural processes may often be described in terms of networks of computational components, such as Boolean logic gates or artificial neurons. The interaction of biological molecules and the flow of information controlling the development and behavior of organisms is particularly amenable to this approach, and these models are well-established in the biological community. However, only relatively recently have papers appeared proposing the use of such systems to perform useful, *human-defined* tasks. For example, rather than merely using the network analogy as a convenient technique for clarifying our understanding of complex systems, it may now be possible to harness the power of such systems for the purposes of computation. In this chapter we review several such proposals, focusing on the molecular implementation of fundamental computational elements. We conclude by describing an instance of cellular computation that has emerged as a result of natural evolution: gene unscrambling in ciliates.

6.1 Introduction

Despite the relatively recent emergence of molecular computing as a distinct research area, the link between biology and computer science is not a new one. Of course, for years biologists have used computers to store and analyze experimental data. Indeed, it is widely accepted that the huge advances of the Human Genome Project (as well as other genome projects) were only made possible by the powerful computational tools available. Bioinformatics has emerged as "the science of the 21st century", requiring the contributions of truly interdisciplinary scientists who are equally at home at the lab bench or writing software at the computer.

However, the seeds of the relationship between biology and computer science were sown over fifty years ago, when the latter discipline did not even exist. When, in the 17th century, the French mathematician and philosopher René Descartes declared to Queen Christina of Sweden that animals could be considered a class of machines, she challenged him to demonstrate how a clock could reproduce. Three centuries later in 1951, with the publication of *"The General and Logical Theory of Automata"* [151] John von Neumann showed how a machine could indeed construct a copy of itself. Von Neumann believed that the behavior of *natural* organisms, although orders of magnitude more complex, was similar to that of the most intricate machines of the day. He believed that life was based on logic.

Twenty years later, the Nobel laureate Jacques Monod identified *specific* natural processes that could be viewed as behaving according to logical principles:

> "The logic of biological regulatory systems abides not by Hegelian laws but, like the workings of computers, by the propositional algebra of George Boole." [109]

This conclusion was drawn from earlier work of Jacob and Monod [110]. In addition, Jacob and Monod described the "lactose system" [82], which is one of the archetypal examples of a Boolean system. We now describe this system in detail.

Genes are composed of a number of distinct regions, which control and encode the desired product. These regions are generally of the form promoter–gene–terminator (Fig. 6.1). Transcription may be regulated by effector molecules known as *inducers* and *repressors*, which interact with the promoter and increase or decrease the level of transcription. This allows effective control over the expression of proteins, avoiding the production of unnecessary compounds. It is important to note at this stage that genetic regulation does not conform to the digital "on-off" model that is popularly portrayed; rather, it is continuous or analog in nature.

One of the most well-studied genetic systems is the *lac operon*. An *operon* is a set of functionally related genes with a common promoter. An example of this is the *lac* operon, which contains three structural genes that allow *E.coli* to utilize the sugar lactose.

When *E.coli* is grown on the common carbon source glucose, the product of the *lacI* gene represses the transcription of the *lacZYA* operon (Fig. 6.2). However, if lactose is supplied *together* with glucose, a lactose by-product is produced which interacts with the repressor molecule, preventing it from repressing the *lacZYA* operon. This de-repression does not itself initiate transcription, since it would be inefficient to utilize lactose if the more common sugar glucose were still available. The operon is positively regulated by the CAP-cAMP (catabolite activator protein: cyclic adenosine monophosphate)

Fig. 6.1. Major regions found within a bacterial operon

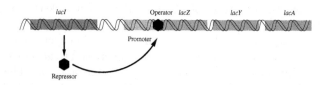

Fig. 6.2. Repressed state of the *lac* operon by the *lacI* gene product

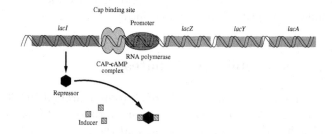

Fig. 6.3. Induced state of the *lac* operon

complex, whose level increases as the amount of available glucose decreases. Therefore, if lactose were present as the sole carbon source, the *lacI* repression would be relaxed and the high CAP-cAMP levels would activate transcription, leading to the synthesis of the *lacZYA* gene products (Fig. 6.3). Thus, the promoter is under the control of two sugars, and the *lacZYA* operon is only transcribed when lactose is present and glucose is absent.

It is clear that we may view the *lac* operon in terms of the Boolean AND function, in that it has output value 1 if transcribed, and 0 otherwise. The presence/absence of glucose corresponds to the gate's first input being equal to 0/1, and the presence/absence of lactose corresponds to the gate's second input being 1/0 (note the difference in representation between the two sugars).

6.2 Successful Implementations

We now briefly introduce several successful examples of how bacteria have been successfully re-engineered for the purposes of computation. In 1999, Weiss et al. [158] described a technique for mapping digital logic circuits onto genetic regulatory networks such that the resulting chemical activity within the cell corresponds to the computations specified by the digital circuit. There was a burst of activity in 2000, when two papers appeared in the same issue of *Nature*, both being seminal contributions to the field. In [57], Elowitz and Leibler described the construction of an oscillator network that periodically caused a culture of *E. coli* to glow by expressing a fluorescent protein. Crucially, the period of oscillation was slower than the cell division cycle, indicating that the state of the oscillator is transmitted from generation to generation. In [62], Gardner et al. implemented a genetic toggle switch in *E. coli*. The switch is flipped from one stable state to another by either chemical or heat induction.

These "single cell" experiments demonstrated the feasibility of implementing artificial logical operations using genetic modification. In [137], Savageau addresses the issue of finding general design principles among microbial genetic circuits, citing several examples. Several more examples of successful work on cellular computing may be found in [7].

6.3 Gene Unscrambling in Ciliates

Ciliate is a term applied to any member of a group of around 10,000 different types of single-celled organism that are characterized by two features: the possession of hair-like *cilia* for movement, and the presence of two kinds of *nuclei* instead of the usual one. One nucleus (the *micronucleus*) is used for sexual exchange of DNA, and the other (the *macronucleus*) is responsible for cell growth and proliferation. Crucially, the DNA in the micronucleus contains an "encoded" description of the DNA in the working macronucleus, which is decoded during development. This encoding "scrambles" fragments of the functional genes in the macronucleus by both the permutation (and possible inversion) of partial *coding* sequences and the inclusion of *non*-coding sequences. A picture of the ciliate *Oxytricha nova* is shown in Fig. 6.4.

It is the macronucleus (that is, the "housekeeping" nucleus) that provides the RNA "blueprints" for the production of proteins. The micronucleus, on the other hand, is a dormant nucleus which is activated only during sexual reproduction, when at some point a micronucleus is converted into a macronucleus in a process known as *gene assembly*. During this process the micronuclear genome is converted into the macronuclear genome. This conversion reorganizes the genetic material in the micronucleus by removing noncoding sequences and placing coding sequences in their correct order. This "unscrambling" may be interpreted as a computational process.

Fig. 6.4. *Oxytricha nova* (Picture courtesy of D.M. Prescott)

The exact mechanism by which genes are unscrambled is not yet fully understood. We first describe experimental observations that have at least suggested possible mechanisms. We then describe two different models of the process. We conclude with a discussion of the computational and biological implications of this work.

6.4 Biological Background

The macronucleus consists of millions of short DNA molecules that result from the conversion of the micronuclear DNA molecules. With few exceptions, each macronuclear molecule corresponds to an individual gene, varying in size between 400 b.p. (*base pairs*) and 15,000 b.p. (the average size is 2000 b.p.). The fragments of macronuclear DNA form a very small proportion of the micronucleus, as up to 98% of micronuclear DNA is noncoding, including intergenic "spacers" (that is, only ∼ 2% of the micronucleus is coding DNA), and all noncoding DNA is excised during gene assembly.

6.4.1 IESs and MDSs

The process of decoding *individual* gene structures is therefore what interests us here. In the simplest case, micronuclear versions of macronuclear genes contain many short, noncoding sequences called *internal eliminated sequences*, or IESs. These are short, AT-rich sequences, and, as their name suggests, they are removed from genes and destroyed during gene assembly. They separate the micronuclear version of a gene into *macronuclear destined sequences*, or MDSs (Fig. 6.5a). When IESs are removed, the MDSs making up a gene are "glued" together to form the functional macronuclear sequence. In the simplest case, IESs are bordered on either side by pairs of identical repeat sequences (pointers) in the ends of the adjacent MDSs (Fig. 6.5b).

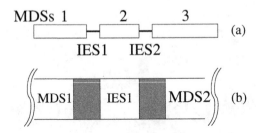

Fig. 6.5. (a) Schematic representation of interruption of MDSs by IESs. (b) Repeat sequences in MDSs flanking an IES (the outgoing repeat sequence of MDS1 is equal to the incoming repeat sequence of MDS2)

6.4.2 Scrambled Genes

In some organisms, the gene assembly problem is complicated by the "scrambling" of MDSs within a particular gene. In this situation, the correct arrangement of MDSs in a macronuclear gene is present in a permuted form in the micronuclear DNA. For example, the actin I gene in *Oxytricha nova* is made up of 9 MDSs and 8 IESs, the MDSs being present in the micronucleus in the order 3–4–6–5–7–9–2–1–8, with MDS2 being inverted [126]. During the development of the macronucleus, the MDSs making up this gene are rearranged into the correct order at the *same* time as IES excision. Scrambling is often *further* complicated by the fact that some MDSs may be *inverted* (a 180° point rotation).

6.4.3 Fundamental Questions

Ciliates are remarkably successful organisms. The range of DNA manipulation and reorganization operations they perform has clearly been acquired during millennia of evolution. However, some fundamental questions remain: what are the underlying molecular mechanisms of gene reconstruction and how did they evolve, and how do ciliates "know" which sequences to remove and which to keep?

Concerning the first question, Prescott proposes [123] that the "compression" of a working nucleus from a larger predecessor is part of a strategy to produce a "streamlined" nucleus in which "every sequence counts" (i.e., useless DNA is not present). This efficiency may be further enhanced by the dispersal of genes into individual molecules, rather than having them being joined into chromosomes. However, so far we still know very little about the details and evolutionary origins of this intricate underlying molecular "machinery."

We may, perhaps, have more success in attempting to answer the second question: how are genes successfully reassembled from an encoded version? In the rest of this chapter we address this question from a computational perspective, and describe two extant models that describe the rearrangement process.

6.5 Models of Gene Construction

We now present a review of two models that attempt to shed light on the process of macronuclear gene assembly. The first model was formulated by Landweber and Kari in [88], where they propose two main operations that model the process of inter- and intramolecular recombination. These can be used to unscramble a micronuclear gene to form a functional copy of the gene in the macronucleus. Both of these operations are based on the concept of repeat sequences "guiding" the unscrambling process.

The first operation takes as input a single linear DNA strand containing two copies of a repeat sequence x. The operation then "folds" the linear strand into a loop, thus aligning the copies of x. The operation then "cuts" the strands in a specific "staggered fashion" within the first copy of x and the second copy of x, creating three strands (ux, wx and v). The operation finally recombines ux and v, and wx forms a circular string. The output of the operation is therefore a linear string and a circular string. This operation mimics the excision of an IES that occurs between two MDSs that are in the correct (i.e., unscrambled) order. In this case the IES is excised as a circular molecule and the two MDSs are "sewn" together to make a single larger MDS.

The second operation takes as input a single linear strand and a separate circular strand. The operation takes two inputs and creates a single linear strand. This allows the insertion of the linear form of the circular strand within the linear strand and mimics intermolecular recombination.

The second model, proposed by Prescott, Ehrenfeucht and Rozenberg (see, for example, [125]), is based on three intramolecular operations (that is, a single molecule folds on itself and swaps part of its sequence through recombination). In this model, we assume full knowledge of the pointer structure of the molecule.

The first operation is the simplest, and is referred to as *loop, direct-repeat excision*. This operation deals with the situation depicted in Fig. 6.6, where two MDSs (x and z) in the correct (i.e., unscrambled) order are separated by an IES, y.

The operation proceeds as follows. The strand is folded into a loop with the two identical pointers aligned (Fig. 6.6a), and then staggered cuts are

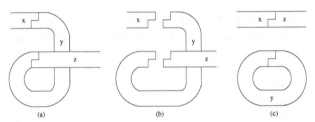

Fig. 6.6. Excision

made (Fig. 6.6b). The pointers connecting the MDSs then join them together, while the IES self-anneals to yield a circular molecule (Fig. 6.6c).

The second operation is known as *hairpin, inverted repeat excision,* and is used in the situation where a pointer has two occurrences, one of which is inverted. The molecule folds into a hairpin structure (Fig. 6.7a) with the pointer and its inversion aligned, cuts are made (Fig. 6.7b) and the inverted sequence is reinserted (Fig. 6.7c), yielding a single molecule.

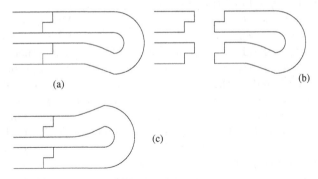

Fig. 6.7. Inversion

The third and final operation is *double-loop, alternating direct repeat excision/reinsertion.* This operation is applicable in situations where two repeats of two pointers have interleaving occurrences on the same strand. The double loop folding is made such that the two pairs of identical pointer occurrences are aligned (Fig. 6.8a), cuts are made (Fig. 6.8b) and the recombination takes place, yielding the molecule from Fig. 6.8c.

The process by which gene assembly takes place using these operations and the computational properties of the system are discussed in detail in [56]. However, the important difference between this model and that of Landweber and Kari is that it is more concerned with the mechanics of the gene assembly process than the computational power of an abstraction of the natural system.

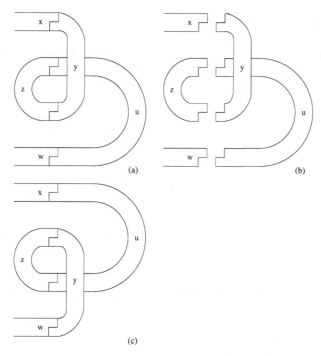

Fig. 6.8. Excision/inversion

The model has been successfully applied to all known experimental data on the assembly of real genes, including the actin I gene of *Urostyla grandis* and *Engelmanniella mobilis*, the gene encoding α telomere binding protein in several stichotrich species, and assembly of the gene encoding DNA polymerase α in *Sterkiella nova*. Descriptions of these applications are presented in [124].

6.6 Summary

In this chapter we have introduced the notion of computing with and inside living cells and reviewed several models for the assembly of genes in ciliates. Although the fundamental molecular mechanisms underlying the operations within these models are still not well-understood, they do suggest possible areas of experimental enquiry. Looking further ahead, it may well be that in the future these mechanisms may even be exploited by using ciliates as prototype cellular computers. This engineering process has already begun, and we have cited several successful examples of the genetic modification of organisms for computational purposes.

6.7 Bibliographical Notes

A comprehensive review of the state of the art in cellular computing is [7]. The definitive text on computation in ciliates is [55], which covers both the biological and theoretical aspects of this area of research.

References

1. Roger L.P. Adams, John T. Knowler, and David P. Leader. *The Biochemistry of the Nucleic Acids.* Chapman and Hall, tenth edition, 1986.
2. Leonard Adleman. Computing with DNA. *Scientific American,* 279(2):54–61, August 1998.
3. Leonard M. Adleman. Molecular computation of solutions to combinatorial problems. *Science,* 266:1021–1024, 1994.
4. Leonard M. Adleman. On constructing a molecular computer. In Baum and Lipton [25].
5. A.V. Aho, J.E. Hopcroft, and J.D. Ullman. *The Design and Analysis of Computer Algorithms.* Addison-Wesley, 1974.
6. Martyn Amos. *DNA Computation.* PhD thesis, Department of Computer Science, University of Warwick, UK, September 1997.
7. Martyn Amos, editor. *Cellular Computing.* Series in Systems Biology. Oxford University Press, USA, 2004.
8. Martyn Amos, Paul E. Dunne, and Alan Gibbons. DNA simulation of Boolean circuits. In John R. Koza, Wolfgang Banzhaf, Kumar Chellapilla, Kalyanmoy Deb, Marco Dorigo, David B. Fogel, Max H. Garzon, David E. Goldberg, Hitoshi Iba, and Rick Riolo, editors, *Genetic Programming 1998: Proceedings of the Third Annual Conference,* pages 679–683. Morgan Kaufmann, July 1998.
9. Martyn Amos, Paul E. Dunne, and Alan Gibbons. Efficient time and volume DNA simulation of CREW PRAM algorithms. Research Report CTAG-98006, Department of Computer Science, University of Liverpool, 1998.
10. Martyn Amos, Alan Gibbons, and Paul E. Dunne. The complexity and viability of DNA computations. In Lundh, Olsson, and Narayanan, editors, *Proceedings of the First International Conference on Bio-computing and Emergent Computation,* pages 165–173, University of Skövde, Sweden, 1997. World Scientific.
11. Martyn Amos, Alan Gibbons, and Paul E. Dunne. Toward feasible and efficient DNA computation. *Complexity,* 4(1):14–18, 1998.
12. Martyn Amos, Alan Gibbons, and David Hodgson. Error-resistant implementation of DNA computations. In Landweber and Baum [94].
13. Martyn Amos, Gheorge Păun, Grzegorz Rozenberg, and Arto Salomaa. Topics in the theory of DNA computing. *Theoretical Computer Science,* 287:3–38, 2002.

14. Martyn Amos, Steve Wilson, David A. Hodgson, Gerald Owenson, and Alan Gibbons. Practical implementation of DNA computations. In C.S. Calude, J. Casti, and M.J. Dinneen, editors, *Unconventional Models of Computation*, Discrete Mathematics and Theoretical Computer Science, pages 1–18. Springer, Singapore, 1998.
15. K. Appel and W. Haken. Every planar map is four colorable. *American Mathematical Society Bulletin*, 82(5):711–712, 1976.
16. K. Appel and W. Haken. *Every Planar Map is Four Colorable*, volume 98 of *Contemporary Mathematics*. American Mathematical Society, 1989.
17. A. Arkin and J. Ross. Computational functions in biochemical reaction networks. *Biophysical Journal*, 67:560–578, 1994.
18. Frederick M. Ausubel, Roger Brent, Robert E. Kingston, David D. Moore, J.G. Seidman, John A. Smith, and Kevin Struhl, editors. *Short Protocols in Molecular Biology*. Wiley, 4th edition, 1999.
19. Eric Bach, Anne Condon, Elton Glaser, and Celena Tanguay. DNA models and algorithms for NP-Complete problems. In *Proceedings of the 11th Conference on Computational Complexity*, pages 290–299. IEEE Computer Society Press, 1996.
20. Jean-Pierre Banâtre and Daniel Le Métayer. The GAMMA model and its discipline of programming. *Science of Computer Programming*, 15:55–77, 1990.
21. Jean-Pierre Banâtre and Daniel Le Métayer. Programming by multiset transformation. *Communications of the ACM*, 36(1):98–111, 1993.
22. K.E. Batcher. Sorting networks and their applications. In *Proc. American Federation of Information Processing Societies 1968 Spring Joint Computer Conference*, volume 32, pages 307–314, Washington, D.C., 1968. Thompson Book Co.
23. Eric B. Baum. DNA sequences useful for computation. In Landweber and Baum [94].
24. Eric B. Baum and Dan Boneh. Running dynamic programming algorithms on a DNA computer. In Landweber and Baum [94].
25. Eric B. Baum and Richard J. Lipton, editors. *DNA Based Computers*, volume 27 of *DIMACS: Series in Discrete Mathematics and Theoretical Computer Science. ISSN: 1052-1798*. American Mathematical Society, 1996.
26. P. Beame, S. Cook, and H. Hoover. Log depth circuits for division and related problems. *SIAM Journal of Computing*, 15(4):994–1003, 1986.
27. Yaakov Benenson, Rivka Adam, Tamar Paz-Livneh, and Ehud Shapiro. DNA molecule provides a computing machine with both data and fuel. *Proceedings of the National Academies of Science*, 100(5):2191–2196, 2003.
28. Yaakov Benenson, Tamar Paz-Elizur, Rivka Adar, Ehud Keinan, Zvi Livneh, and Ehud Shapiro. Programmable and autonomous computing machine made of biomolecules. *Nature*, 414:430–434, 2001.
29. C.H. Bennett. The thermodynamics of computation – a review. *International Journal of Theoretical Physics*, 21:905–940, 1982.
30. Gérard Berry and Gérard Boudol. The chemical abstract machine. *Theoretical Computer Science*, 96:217–248, 1992.
31. New England Biolabs. Catalog, 1996.
32. Dan Boneh, Christopher Dunworth, Richard J. Lipton, and Jiří Sgall. Making DNA computers error resistant. In Landweber and Baum [94].

33. Ravinderjit S. Braich, Nickolas Chelyapov, Cliff Johnson, Paul W.K. Rothemund, and Leonard Adleman. Solution of a 20-variable 3-sat problem on a DNA computer. *Science*, 296:499–502, 2002.

34. Dennis Bray. Protein molecules as computational elements in living cells. *Nature*, 376:307–312, 1995.

35. K.J. Breslauer, R. Frank, H. Blocker, and L.A. Marky. Predicting DNA duplex stability from the base sequence. *Proc. Natl. Acad. Sci.*, pages 3746–3750, 1986.

36. T.A. Brown. *Gene Cloning: an Introduction.* Chapman and Hall, second edition, 1990.

37. T.A. Brown. *Genetics: A Molecular Approach.* Chapman and Hall, 1993.

38. B. Bunow. On the potential of molecular computing. *Science*, 268:482–483, April 1995.

39. Cristian S. Calude and Gheorghe Paun. *Computing with Cells and Atoms: An Introduction to Quantum, DNA and Membrane Computing.* Taylor and Francis, 2001.

40. Scott Camazine, Jean-Louis Deneubourg, Nigel R. Franks, James Sneyd, Guy Theraulaz, and Eric Bonabeau. *Self-organization in Biological Systems.* Princeton University Press, 2001.

41. M. Chandy and J. Misra. *Parallel Programming Design: A Foundation.* Addison-Wesley, 1988.

42. Junghuei Chen, Russell Deaton, and Yu-Zhen Wang. A DNA-based memory with *in vitro* learning and associative recall. In Chen and Reif [43]. LNCS 2943.

43. Junghuei Chen and John Reif, editors. *Proceedings of the 9th International Workshop on DNA Based Computers.* Springer, 2003. LNCS 2943.

44. Junghuei Chen and David Harlan Wood. Computation with biomolecules. *Proc. Natl. Acad. Sci.*, 97(4):1328–1330, 2000.

45. Michael Conrad. On design principles for a molecular computer. *Communications of the ACM*, 28:464–480, 1985.

46. Michael Conrad and E.A. Liberman. Molecular computing as a link between biological and physical theory. *Journal of Theoretical Biology*, 98:239–252, 1982.

47. Stephen Cook. The complexity of theorem proving procedures. In *Proceedings of the Third Annual ACM Symposium on Theory of Computing*, pages 151–158, 1971.

48. T. Cormen, C. Leiserson, R. Rivest, and C. Stein. *Introduction to Algorithms.* MIT Press, 2nd edition, 2001.

49. Francis Crick. Central dogma of molecular biology. *Nature*, 227:561–563, 1970.

50. Erzsébet Csuhaj-Varjú, R. Freund, Lila Kari, and Gheorghe Păun. DNA computation based on splicing: universality results. In Lawrence Hunter and Teri Klein, editors, *Biocomputing: Proceedings of the 1996 Pacific Symposium.* World Scientific, January 1996.

51. R. Deaton, Max H. Garzon, R.C. Murphy, Donald R. Franceschetti, and S.E. Stevens, Jr. Genetic search of reliable encodings for DNA based computation. In *First Conference on Genetic Programming*, Stanford University, 1996.

52. R. Deaton, R.C. Murphy, M. Garzon, D.R. Franceschetti, and S.E. Stevens, Jr. Good encodings for DNA-based solutions to combinatorial problems. In Landweber and Baum [94].

53. Dónall A. Mac Dónaill. On the scalability of molecular computational solutions to NP problems. *The Journal of Universal Computer Science*, 2(2):87–95, February 1996.

54. Paul E. Dunne. *The Complexity of Boolean Networks*. Academic Press, 1988.

55. A. Ehrenfeucht, T. Harju, I. Petre, D.M. Prescott, and G. Rozenberg. *Computation in Living Cells: Gene Assembly in Ciliates*. Springer, 2004.

56. A. Ehrenfeucht, I. Petre, D.M. Prescott, and G. Rozenberg. Formal systems for gene assembly in ciliates. *Theoretical Computer Science*, 292(1):199–219, 2003.

57. M. Elowitz and S. Leibler. A synthetic oscillatory network of transcriptional regulators. *Nature*, 403:335–338, January 2000.

58. Dirk Faulhammer, Anthony R. Cukras, Richard J. Lipton, and Laura F. Landweber. Molecular computation: RNA solutions to chess problems. *Proceedings of the National Academies of Science*, 97(4):1385–1389, 2000.

59. Richard P. Feynman. There's plenty of room at the bottom. In D. Gilbert, editor, *Miniaturization*, pages 282–296. Reinhold, 1961.

60. M. Fischer and N.J. Pippenger. Relations among complexity measures. *Journal of the ACM*, 26:361–381, 1979.

61. Steven Fortune and James Wyllie. Parallelism in random access machines. In *Conference Record of the Tenth Annual ACM Symposium on Theory of Computing*, pages 114–118, San Diego, California, 1978.

62. T. Gardner, R. Cantor, and J. Collins. Construction of a genetic toggle switch in *Escherichia coli*. *Nature*, 403:339–342, January 2000.

63. Michael R. Garey and David S. Johnson. *Computers and Intractability: A Guide to the Theory of NP-Completeness*. W.H. Freeman and Company, New York, 1979.

64. A. Gibbons and W. Rytter. *Efficient Parallel Algorithms*. Cambridge University Press, 1988.

65. Alan Gibbons, Martyn Amos, and David Hodgson. Models of DNA computation. In Penczek and Szalas, editors, *Mathematical Foundations of Computer Science (MFCS)*, volume 1113 of *Lecture Notes in Computer Science*, pages 18–36. Springer, 1996.

66. Alan Gibbons and Paul Spirakis, editors. *Lectures in Parallel Computation*. Cambridge University Press, 1993.

67. A.M. Gibbons. *Algorithmic Graph Theory*. Cambridge University Press, 1985.

68. D.E. Goldberg. *Genetic Algorithms in Search, Optimization, and Machine Learning*. Addison-Wesley, 1989.

69. Larry Gonick and Mark Wheelis. *The Cartoon Guide to Genetics*. Harper-Perennial, 1983.

70. John Gribbin. *In Search of Schroedinger's Cat: Quantum Physics and Reality*. Corgi, 1984.

71. Frank Guarnieri, Makiko Fliss, and Carter Bancroft. Making DNA add. *Science*, 273:220–223, 1996.

72. Z. Guo, R.A Guilfoyle, A.J Thiel, R. Wang, and L.M Smith. Direct fluorescence analysis of genetic polymorphisms by hybridization with oligonucleotide arrays on glass supports. *Nucl. Acids Res.*, 22:5456–5465, 1994.

73. Masami Hagiya and Azuma Ohuchi, editors. *Proceedings of the 8th International Workshop on DNA Based Computers*. Springer, 2002. LNCS 2568.

74. Andrew Hamilton. Brains that click. *Popular Mechanics*, 91(3):162–167,256,258, March 1949.

75. M.A. Harrison. *Introduction to Switching and Automata Theory.* McGraw-Hill, 1965.
76. Juris Hartmanis. On the weight of computations. *Bulletin of the European Association for Theoretical Computer Science,* 55:136–138, 1995.
77. Thomas Head. Formal language theory and DNA: an analysis of the generative capacity of specific recombinant behaviors. *Bulletin of Mathematical Biology,* 49(6):737–759, 1987.
78. Thomas Head. Splicing schemes and DNA. In G. Rozenberg and A. Salomaa, editors, *Lindenmayer Systems: Impacts on Theoretical Computer Science, Computer Graphics and Developmental Biology,* pages 371–384. Springer, 1992.
79. A. Hjelmfelt, F.W. Schneider, and J. Ross. Pattern recognition in coupled chemical kinetic systems. *Science,* pages 335–337, 1993.
80. Allen Hjelmfelt, Edward D. Weinberger, and John Ross. Chemical implementation of neural networks and Turing machines. *Proceedings of the National Academy of Sciences,* 88:10983–10987, 1991.
81. D.A. Jackson, R.H. Symons, and P. Berg. Biochemical method for inserting new genetic information into DNA of simian virus 40: circular SV40 DNA molecules containing lambda phage genes and the galactose operon of *Escherichia coli. Proc. Natl. Acad. Sci. USA,* 69:2904–2909, 1972.
82. F. Jacob and J. Monod. Genetic regulatory mechanisms in the synthesis of proteins. *Journal of Molecular Biology,* 3:318–356, 1961.
83. Joseph JáJá. *An Introduction to Parallel Algorithms.* Addison-Wesley, 1992.
84. Natasha Jonoska, Gheorge Păun, and Grzegorz Rozenberg, editors. *Aspects of Molecular Computing: Essays Dedicated to Tom Head on the Occasion of his 70th Birthday.* Springer, 2004. Lecture Notes in Computer Science Volume 2950.
85. P.D. Kaplan, Q. Ouyang, D.S. Thaler, and A. Libchaber. Parallel overlap assembly for the construction of computational DNA libraries. *Journal of Theoretical Biology,* 188(3):333–341, 1997.
86. Peter Kaplan, Guillermo Cecchi, and Albert Libchaber. *Molecular Computation: Adleman's Experiment Repeated.* Technical report, NEC Research Institute, 1995.
87. Peter D. Kaplan, Guillermo Cecchi, and Albert Libchaber. DNA based molecular computation: template-template interactions in PCR. In Landweber and Baum [94].
88. Lila Kari and Laura F. Landweber. Computational power of gene rearrangement. In Erik Winfree and David K. Gifford, editors, *Proceedings 5th DIMACS Workshop on DNA Based Computers, held at the Massachusetts Institute of Technology, Cambridge, MA, USA June 14–June 15, 1999,* pages 207–216. American Mathematical Society, 1999.
89. Richard M. Karp. Reducibility among combinatorial problems. In Raymond E. Miller and James W. Thatcher, editors, *Complexity of Computer Computations,* pages 85–103. Plenum Press, 1972.
90. Richard M. Karp, Claire Kenyon, and Orli Waarts. Error-resilient DNA computation. In *7th ACM-SIAM Symposium on Discrete Algorithms,* pages 458–467. SIAM, 1996.
91. V.M. Khrapchenko. Asymptotic estimation of addition time of a parallel adder. *Prob. Kibernet.,* 19:107–122, 1967. In Russian, translation in *Syst. Theory. Res.,* 19, (1970), 105–122.

92. D. Knuth. *The Art of Computer Programming: Sorting and Searching*, volume 3. Addison-Wesley, second edition, 1998.

93. Stuart A. Kurtz, Stephen R. Mahaney, James S. Royer, and Janos Simon. Active transport in biological computing. In Landweber and Baum [94].

94. Laura F. Landweber and Eric B. Baum, editors. *Proceedings of the Second Annual Meeting on DNA Based Computers, DIMACS: Series in Discrete Mathematics and Theoretical Computer Science*. American Mathematical Society, June 1996.

95. L.F. Landweber and R.J. Lipton. DNA^2DNA: A potential 'killer app'? In *24th International Colloquium on Automata, Languages and Programming (ICALP)*, Lecture Notes in Computer Science, pages 672–683. Springer, 1997.

96. G.E. Liepins and M.D. Vose. Characterizing crossover in genetic algorithms. *Annals of Mathematics and Artificial Intelligence*, 5(1):27–34, April 1992.

97. M. Linial and N. Linial. On the potential of molecular computing. *Science*, 268:481, April 1995.

98. Richard J. Lipton. DNA solution of hard computational problems. *Science*, 268:542–545, 1995.

99. Qinghua Liu, Liman Wang, Anthony G. Frutos, Anne E. Condon, Robert M. Corn, and Lloyd M. Smith. DNA computing on surfaces. *Nature*, 403:175–179, 2000.

100. Quinghua Liu, Zhen Guo, Anne E. Condon, Robert M. Corn, Max G. Lagally, and Lloyd M. Smith. A surface-based approach to DNA computation. In Landweber and Baum [94].

101. Y.M.D. Lo, K.F.C. Yiu, and S.L. Wong. On the potential of molecular computing. *Science*, 268:481–482, April 1995.

102. P.E. Lobban and C.A. Sutton. Enzymatic end-to-end joining of DNA molecules. *J. Mol. Biol.*, pages 453–471, 1973.

103. Chengde Mao, Thomas H. LaBean, John H. Reif, and Nadrian C. Seeman. Logical computation using algorithmic self-assembly of DNA triple-crossover molecules. *Nature*, 407:493–496, 2000.

104. J. Marmur. A procedure for the isolation of deoxyribonucleic acid from microorganisms. *Journal of Molecular Biology*, 3:208–218, 1961.

105. Daisuke Matsuda and Masayuki Yamamura. Cascading whiplash PCR with a nicking enzyme. In Hagiya and Ohuchi [73]. LNCS 2568.

106. Carver Mead and Lynn Conway. *Introduction to VLSI systems*. Addison-Wesley, 1980.

107. Zbigniew Michalewicz and David B. Fogel. *How to Solve It: Modern Heuristics*. Springer, 2000.

108. Kalim U. Mir. A restricted genetic alphabet for DNA computing. In Landweber and Baum [94].

109. Jacques Monod. *Chance and Necessity*. Penguin, 1970.

110. Jacques Monod, Jean-Pierre Changeux, and Francois Jacob. Allosteric proteins and cellular control systems. *Journal of Molecular Biology*, 6:306–329, 1963.

111. Kary B. Mullis. The unusual origin of the polymerase chain reaction. *Scientific American*, 262:36–43, 1990.

112. Kary B. Mullis, François Ferré, and Richard A. Gibbs, editors. *The Polymerase Chain Reaction*. Birkhauser, 1994.

113. Mitsunori Ogihara and Animesh Ray. Simulating Boolean circuits on a DNA computer. In *Proceedings of the First Annual International Conference on*

Computational Molecular Biology (RECOMB97), pages 226–231. ACM Press, August 1997.

114. R. Old and S. Primrose. *Principles of Gene Manipulation, an Introduction to Genetic Engineering*. Blackwell, fifth edition, 1994.

115. Qi Ouyang, Peter D. Kaplan, Shumao Liu, and Albert Libchaber. DNA solution of the maximal clique problem. *Science*, 278:446–449, 1997.

116. G. Paun. Computing with membranes (P systems): a variant. *International Journal of Foundations of Computer Science (IJFCS)*, 11(1):167–182, 2000.

117. G. Paun. Computing with membranes (P systems): Twenty-six research topics. Report CDMTCS-119, Centre for Discrete Mathematics and Theoretical Computer Science, University of Auckland, Auckland, New Zealand, February 2000.

118. Gheorghe Păun. Computing with membranes. *Journal of Computer and System Sciences*, 61(1):108–143, 2000.

119. Gheorghe Păun. *Membrane Computing: An Introduction*. Springer, 2002.

120. Gheorghe Păun, Grzegorz Rozenberg, and Arto Salomaa. *DNA Computing: New Computing Paradigms*. Springer, 1998.

121. M.F. Perutz, S.S. Hasnain, P.J. Duke, and J.L. Sessler. Stereochemistry of iron in deoxyhaemoglobin. *Nature*, 295:535, 1982.

122. Richard Powers. *The Gold Bug Variations*, page 369. HarperPerennial, 1991.

123. David M. Prescott. Invention and mystery in hypotrich DNA. *Journal of Eukaryotic Microbiology*, 45(6):575–581, 1998.

124. David M. Prescott and Grzegorz Rozenberg. Encrypted genes and their assembly in ciliates. In Amos [7].

125. D.M. Prescott, A. Ehrenfeucht, and G. Rozenberg. Molecular operatons for DNA processing in hypotrichous ciliates. *European Journal of Protistology*, 37:241–260, 2001.

126. D.M. Prescott and A.F. Greslin. Scrambled actin I gene in the micronucleus of *Oxytricha nova*. *Developmental Genetics*, 13:66–74, 1992.

127. C. Queen and L.J. Korn. A comprehensive sequence analysis programme for the IBM personal computer. *Nucleic Acids Research*, 12(1):581–599, 1984.

128. John H. Reif. Parallel molecular computation: Models and simulations. In *Proceedings of the Seventh Annual ACM Symposium on Parallel Algorithms and Architectures (SPAA95), Santa Barbara, June 1995*, pages 213–223. Association for Computing Machinery, June 1995.

129. W. Reisig and G. Rozenberg, editors. *Lectures on Petri Nets I: Basic Models*. Springer, 1998.

130. L. Roberts and C. Murrell, editors. *Understanding Genetic Engineering*. Ellis Horwood, 1989.

131. Lynne Roberts and Colin Murrell (Eds.). An introduction to genetic engineering. Department of Biological Sciences, University of Warwick, 1998.

132. Paul W.K. Rothemund and Erik Winfree. The program-size complexity of self-assembled squares (extended abstract). In *Proceedings of the Thirty-Second Annual ACM Symposium on Theory of Computing*, pages 459–468. ACM Press, 1999.

133. Sam Roweis, Erik Winfree, Richard Burgoyne, Nickolas V. Chelyapov, Myron F. Goodman, Paul W.K. Rothemund, and Leonard M. Adleman. A sticker based architecture for DNA computation. In Landweber and Baum [94].

134. Harvey Rubin. Looking for the DNA killer app. *Nature Structural Biology*, 3(8):656–658, August 1996. Editorial.

135. Kensaku Sakamoto, Hidetaka Gouzu, Ken Komiya, Daisuke Kiga, Shigeyuki Yokoyama, Takashi Yokomori, and Masami Hagiya. Molecular computation by DNA hairpin formation. *Science*, 288:1223–1226, 2000.
136. J. Sambrook, E.F. Fritsch, and T. Maniatis. *Molecular Cloning: A Laboratory Manual.* Cold Spring Harbor Press, second edition, 1989.
137. Michael A. Savageau. Design principles for elementary gene circuits: Elements, methods and examples. *Chaos*, 11(1):142–159, 2001.
138. Roger Sayle and E. James Milner-White. Rasmol: biomolecular graphics for all. *Trends in Biochemical Sciences (TIBS)*, 20(9):374, September 1995.
139. C.P. Schnorr. The network complexity and Turing machine complexity of finite functions. *Acta Informatica*, 7:95–107, 1976.
140. A. Schonhage and V. Strassen. Schnelle multiplikation grosser zahlen. *Computing*, 7:281–292, 1971.
141. Nadrian C. Seeman, Hui Wang, Bing Liu, Jing Qi, Xiaojun Li, Xiaoping Yang, Furong Liu, Weiqiong Sun, Zhiyong Shen, Ruojie Sha, Chengde Mao, Yinli Wang, Siwei Zhang, Tsu-Ju Fu, Shouming Du, John E. Mueller, Yuwen Zhang, and Junghuei Chen. The perils of polynucleotides: the experimental gap between the design and assembly of unusual DNA structures. In Landweber and Baum [94].
142. N.C. Seeman. DNA nanotechnology: novel DNA constructions. *Annual Review of Biophysics and Biomolecular Structure*, 27:225–248, 1998.
143. H.M. Sheffer. A set of five independent postulates for Boolean algebras, with application to logical constants. *Transactions of the American Mathematical Society*, 14:481–488, 1913.
144. Tanya Sienko, Andrew Adamatzky, and Nicholas Rambidi, editors. *Molecular Computing.* MIT Press, 2003.
145. Michael Sipser. *Introduction to the Theory of Computation.* PWS Publishing, 1997.
146. S. Skiena. *The Algorithm Design Manual.* Springer, 1997.
147. Milan N. Stojanovic and Darko Stefanovic. A deoxribozyme-based molecular automaton. *Nature Biotechnology*, 21(9):1069–1074, 2003.
148. H. Stubbe. *History of Genetics—from Prehistoric times to the Rediscovery of Mendel's Laws.* MIT Press, 1972.
149. Andrew S. Tanenbaum. *Structured Computer Organization.* PrenticeHall, fourth edition, 1999.
150. A. Turing. On computable numbers, with an application to the entscheidungsproblem. *Proceedings of the London Mathematical Society, 2nd series*, 42:230–265, 1936.
151. John von Neumann. The general and logical theory of automata. In *Cerebral Mechanisms in Behavior*, pages 1–41. Wiley, New York, 1941.
152. H. Wang. Proving theorems by pattern recognition. *Bell System Technical Journal*, 40:1–42, 1961.
153. J.D. Watson and F.H.C. Crick. Genetical implications of the structure of deoxyribose nucleic acid. *Nature*, 171:964, 1953.
154. J.D. Watson and F.H.C. Crick. Molecular structure of nucleic acids: a structure for deoxyribose nucleic acid. *Nature*, 171:737–738, 1953.
155. J.D. Watson, M. Gilman, J. Witkowski, and M. Zoller. *Recombinant DNA.* Scientific American Books, second edition, 1992.
156. J.D. Watson, N.H. Hopkins, J.W. Roberts, J.A. Steitz, and A.M. Weiner. *Molecular Biology of the Gene.* BenjaminCummings, fourth edition, 1987.

157. Ingo Wegener. *The Complexity of Boolean Functions*. Wiley-Teubner, 1987.
158. R. Weiss, G. Homsy, and T.F. Knight Jr. Toward in-vivo digital circuits. In L.F. Landweber and E. Winfree, editors, *DIMACS Workshop on Evolution as Computation*. Springer, January 1999.
159. N.E. Weste and K. Eshragan. *Principles of CMOS VLSI Design*. Addison-Wesley, 1993.
160. J. Williams, A. Ceccarelli, and N. Spurr. *Genetic Engineering*. βios Scientific Publishers, 1993.
161. E. Winfree, F. Liu, L. Wenzler, and N.C. Seeman. Design and self-assembly of two-dimensional DNA crystals. *Nature*, 394:539–544, 1998.
162. Erik Winfree. On the computational power of DNA annealing and ligation. In Baum and Lipton [25], pages 199–221.
163. Erik Winfree. *Algorithmic self-assembly of DNA*. PhD thesis, California Institute of Technology, May 1998.
164. C. Yanisch-Perron, J. Vieira, and J. Messing. Improved m13 phage cloning vectors and host strains: nucleotide sequences of the m13mp18 and puc19 vectors. *Gene*, 33(1):103–119, 1985.
165. Claudio Zandron, Claudio Ferretti, and Giancarlo Mauri. Solving NP-complete problems using P systems with active membranes. In I. Antoniou, C.S. Calude, and M.J. Dinneen, editors, *Unconventional Models of Computation, UMC'2K, Solvay Institutes, Brussel, 13–16 December 2000*, DIMACS: Series in Discrete Mathematics and Theoretical Computer Science, pages 289–301. Centre for Discrete Mathematics and Theoretical Computer Science, International Solvay Institute for Physics and Chemistry, and the Vrije Universiteit Brussel, Theoretical Physics Division, 2000.

Index

ALGOL, 102
AND, 23–24, 42, 87–88, 93, 99, 149
accumulator, 30–31, 93, 97–99
actin I, 152, 155
active site, 8, 110
addition, 29, 93, 98, 143–144
adenine, 6, 131
adjacency matrix, 34, 89
agarose, 13, 114, 125, 131
algorithm,
alphabet, 26, 47, 55, 60, 66–67, 73, 83
amino acid, 8
annealing, 6, 11, 16, 86, 111, 119, 126,
    128, 131, 133, 134, 139
architecture, 1-3, 23, 44, 64, 102, 140
array, 33–34, 44, 64, 140, 144
automation, 15, 120, 138, 140
automata, see automaton
automaton, 25–28, 44, 63, 145, 148

backbone, 6, 129, 145
bacteria, 4, 10, 15, 18–20, 125, 150
bacteriophage, see phage
basis, 77
base, 6, 120, 128, 141, 143
base pair, 6, 14, 151
Batcher sorting network, see network,
    sorting
Bennett, C., 109
big-oh notation, 41
binary, 23, 55, 62–63, 97–98, 142–144
bioinformatics, 8, 147
biology (as a discipline), 2–4, 147–148
biotechnology, 4

biotin, 12
biotinylation, 111, 121–122, 129–131,
    135
bit, 23, 49, 55, 89, 97–99, 140, 142–144
blunt end, 16, 84, 86
bonding, see annealing
Boolean
    algebra 24–25, 44, 148
    circuit, see circuit, Boolean
    matrix 89
    value 49, 89, 97
    variable 55, 77, 82–83, 97
bubble sort, 41, 43

calcium, 18
capsid, 18
cardinality, 46, 110
catalyst, 7–8
cell (biological), 2, 17–19, 66, 150
cellular automaton, 63
central dogma of molecular biology, 7
central processing unit (CPU), 1–2
charge, 13, 68
CHemical Abstract Machine (CHAM),
    see model, CHAM
chess game, 136
chromosome, 5, 153
Church's Thesis, 29
circuit
    integrated, 1
    Boolean 24–25, 41–42, 71, 73, 77–83,
        86–91, 94–100, 102, 105–107, 145,
        150
    genetic 150

clause, 42, 49, 139–141, 144
clique, 39, 142–143
clock, 67, 148
cloning, 17–18, 21, 118, 125–126,
    129–130, 134
coloring, 37, 46–48, 52, 120–122,
    124–126, 130
combinatorial explosion, 42
complement
    arithmetic, 98
    graph 142–143
complementarity (biological), 6–7,
    10–12, 14, 17, 79, 83–84, 86,
    110–111, 121, 134, 138–139, 144
complex system, 147
complexity 3, 5, 64, 71
complexity
    of Boolean circuit, 78, 83
    computational, 39–41, 44–45, 73,
        75–77, 93, 107
    class, 43–44, 50–51, 60, 72, 94, 107,
        115
computational problem, 3, 33, 73, 140
computer, 1–4, 23, 26, 30, 33, 40–41,
    43–44, 141, 147–148
    Brownian, 109
    cellular, 155
    DNA-based, 2, 140–141
computer science, 3–4, 23, 27, 43, 63,
    147–148
concurrent programming, 64
conformation, 8, 110–111
connectivity, 36
Conrad, M., 109–110
control, 2, 32, 61, 64, 66, 68, 92, 110
    experimental, 114, 116, 122, 124,
        126–131, 133–134
    genetic, 147–149
    graph, 95–97
    program, 92–95, 100, 103
Cook's Theorem, 43
cytosine, 6

DNA
    charge, 13
    cloning, see cloning
    as genetic material, 5, 7
    operations on, 10–17
    recombinant, 17

replication, see polymerase chain
    reaction
    structure, 5–6, 17
data, 1–2, 5, 34, 43, 97, 147, 155
    input, 92–93
    structure, 26, 33–34, 44
decision problem, 34, 43
denaturing, 11, 15, 85, 126–128
deoxyribonucleic acid, see DNA
Descartes, R., 148
double helix, see DNA, structure

E. coli, 18, 134, 139, 148, 150
electric field, 13
electroporation, 19
endonuclease, 16, 110
engineering (as a discipline), 4
    genetic, 4, 21, 155
    reverse, 4
enzyme, 7–8, 10, 110
    allosteric, 110
    hypothetical, 109
    ligase, 7, 11, 85–86, 122, 124
    polymerase, 2, 10
    restriction, 16–17, 76, 80, 84–86, 110,
        117–119, 127, 131, 133–134, 136,
        143
        effectiveness, 135
        universal, 136
    RsaI, 16
    Sau3A, 17, 126–127, 130–131,
        134–135
    Taq, 15, 126–128, 131
error
    experimental, 3, 46, 81, 114–116,
        118–120, 135
    overflow, 98–99
    sequence design, 120
ethidium bromide, 13
exonuclease, 116, 139
exponential curse, 72, 115
expression
    genetic, 10, 148
    of algorithm, 31, 51, 55

factor 42, 45, 100
    constant, 41, 81, 96
    factoring algorithm, 40–41
feasibility 3, 88

biological, 74, 114, 150
Feynman, R., 2
function 23–26, 29, 31–32, 42, 65, 77,
    93–94, 99–100
  majority, 78
  of protein, 7–8, 14
  uncomputable, 29

GAMMA, *see* model, GAMMA
gate 4, 23–25, 43, 77–87, 89–90, 92,
    98–100
  biological, 110, 145, 149
  as metaphor, 149
gel, 131
  artifact, 130
  gel-based computing, 140–141
  electrophoresis, 13, 58, 80, 85–86,
    114, 118, 143
  extraction, 126
  purification, 125–126, 131
  visualization, 13–14, 86
gene, 5, 7, 10, 16–17, 151
  assembly, 150–154
  unscrambling, 150
  structural, 10
  structure, 148, 151
genetic algorithm, 60
genome, 4, 21, 147, 150
glucose, 8, 148–149
graph, 34–37, 39, 44, 46–48, 52, 54–55,
    89, 112, 114, 116, 134, 141, 143
  acyclic, 77, 81–82, 95
  branch-free, 96
  complementary, 143
  complete, 55
  control, 95–97
  directed, 88–89
  planar, 44
  theory, 51
guanine, 6

hairpin, 120, 134, 144–145, 154
Halting Problem, 29
Hamiltonian Path Problem, 3, 36, 47,
    52–54, 69, 72, 112–116
Hartmanis, J., 71–72, 115
Heisenberg's Uncertainty Principle, 1
hemoglobin, 8
heuristic, 72

hexokinase, 8
hybridization, 116, 118–119, 121, 124,
    126, 136, 145
hydrogen bond, 6, 10–11, 131

indirect addressing, 31, 98
inducer, 148
infection cycle, 19
information, 2, 23, 33, 44
  biological, 147
  genetic, 5–8, 10
  processing, 2, 110
  storage, 2
instruction, 1–2, 30, 40–41, 92–97, 100
  set, 30–31, 33, 91–92, 102–103
integrated circuit, 1
internal eliminated sequence (IES),
    151–154
inversion, 150, 154

Jacob, F., 10, 148

keratin, 8
killer application, 71–73, 106–107
Klenow, 127, 131
Knight Problem, *see* chess game

lactose, 148–149
lane, 13–14, 114, 131–132
library, 56–57, 59, 75–76, 120, 124, 130,
    134–137, 141, 143
ligand, 110
ligase, *see* enzyme, ligase
ligation, 7, 11, 18, 79–80, 86, 112, 134
list, 26, 33
  ranking, 102–106
literal, 48–49, 81, 144
lock-key, 8, 86, 110
logic, *see* Boolean, algebra

macronuclear destined sequence (MDS),
    151–154
macronucleus, 150–153
magnetic bead separation, 12
majority function, *see* function,
    majority
marker lane, 13–14
mathematics, 4
maximal clique, 141–142
maximum clique, 39, 55

maximum independent set, 55
melting temperature 120, 134
membrane, 66
  cell, 18–19
  elementary, 66
memory, 1–3, 26–27, 30–31, 33, 40,
    91–92, 94–100, 102–103, 105–106
  DNA-based, 145
  strand, 55–59
Mendel, G., 5
metabolism, 7
Microgenie, 120
micronucleus, 150–152
migration rate, 13
miniaturization, 1
minimal set cover, 57
model, 3, 24, 36, 45
  Boolean circuit, 77–88, 91
  CHAM, 64–65
  computational, 4, 33, 73
  constructive, 61
  double helix, 5
  filtering, 46, 110
  GAMMA, 65
  of gene construction, 153–155
  of genetic regulation, 148
  implementation issues, 45–46
  mark and destroy, 3
  membrane, 63, 66
  P-RAM, 72, 90–106
  P system, 66–69
  parallel associative memory (PAM),
    60–61
  parallel filtering, 50–55, 74–75,
    116–119, 137–138, 145
  RAM, 30–33, 63
  satisfiability, 48–49
  splicing, 60
  sticker, 55–60, 88, 140
  strong, 71, 76, 88, 91, 107
  tile assembly, 61–63
  unrestricted, 47–48
  von Neumann, 3
Monod, J., 10, 148
mRNA, see RNA, messenger
multi-set, 46–47, 60, 63, 67, 110–111,
    115

NAND, 77, 82–83, 86–87, 89–90, 92,
    100, 102
NC (complexity class), 72, 91, 94, 100,
    102, 107
NOT, 23–25, 42
NP (complexity class), 43–44, 47–48,
    50–51, 55, 60, 69, 72, 74, 115, 135,
    141
nanotechnology, 2
network, 33–34, 43
  biological, 147
  Boolean, see circuit, Boolean
  combinatorial, 82, 91–92
  depth, 83, 91
  division, 99
  gel, 13
  genetic, 150
  as metaphor, 147
  oscillator, 150
  process, 64
  size, 83, 91
  sorting, 44, 87–88
  telecommunication, 36–37
neuron, 110, 147
noughts and crosses, see tic-tac-toe
nucleotide, 6, 11, 14, 137

OR, 23–24, 42, 87
oligo (oligonucleotide), 6, 11–12, 14, 86,
    110–114, 120–122, 124, 128, 130,
    138–140, 145
operation set, 45, 49–50, 74
operon, 148–149
optical computing, 2
optimisation, 34
orientation, 6
oscillator, 150
Oxytricha nova, 150–152

P (complexity class), 43
P system, see model, P system
parallel random access machine
    (P-RAM), see model, P-RAM
parallel associative matching (PAM),
    see model, parallel associative
    matching
parallelism, 72
permutation, 46, 50–51, 53–54, 74–76,
    150

Petri net, 63
phage, 18–20
phosphate, 6
    monophosphate, 148
    triphosphate, 14
polarity, 80
polyacrylamide, 13, 80, 140
polymer, 6
polymerase, 2–3, 10, 14–16, 117, 124, 155
polymerase chain reaction (PCR), 14–16, 86, 114, 116, 118, 122, 124–131, 133–135, 137, 140–141
    whiplash, 145
prime number, 40, 65
primer, 14–16, 76, 114, 117, 119, 122, 124–128, 131–132, 134–135
processor, 1–2, 52, 72–73, 91–95, 97, 100, 102–106
program, 26–32, 39–40, 64–65, 91–92, 94–97, 100, 102
program counter (PC), 30
promoter, 10, 148–149
protein, 2, 4–5, 7–10, 18, 20–21, 62, 145, 148, 150, 155
pseudo-code, 31–32, 40, 49

quantum computing, 2
quantum effect, 1
quicksort, 41

random access machine (RAM), see model, RAM
random search, 2, 143
reaction, 7–8, 50, 65, 76, 109–110, 127, 143
    condition, 65
    cross, 129–130
    detection, 130
    dynamics,
    horizontal chain, 144
    kinetics,
    rule, 66
    sequencing, 129
recombinant DNA, see DNA, recombinant
repression, 148–149
repressor, 148

resource, 3, 33, 39–40, 44, 71–73, 90, 106–107
restriction enzyme, see enzyme, restriction
restriction site, 16–17, 79, 84–86, 110, 117, 119–120, 143
ribonuclease, 136, 138
RNA, 7, 10, 109, 136–138, 150
    messenger, 10
robot, 106
RsaI, see enzyme, RsaI
running time, 33, 40–41, 72–73, 94, 100

salt, 133
satisfiability (SAT), 41, 48–50, 136, 138–140, 144
Sau3A, see enzyme, Sau3A
scalability, 115–116, 140, 143–145
scheduling, 43
self-assembly, 8, 61–63, 144–145
separation, 11–12, 50, 75, 118–119, 140
sequence design, 11, 14, 79–80, 84–85, 110, 120, 128, 134–135, 141
shock, 18
specificity, 10, 126–127, 134
state, 1, 25–28, 41, 60–62, 92, 94, 109–110, 150
    stable, 65
sticker, see model, sticker
sticky end, 17–18, 63, 144
strand
    memory, see memory, strand
    rule, 63
string, 26–29, 46–48, 50–54, 60–61, 67, 73–76, 78–79, 83–84, 137, 142–144, 153
subgraph, 37, 39, 54–55, 141–142
subgraph isomorphism, 54–55
substrate, 8, 45, 110
    computational, 1, 21, 73, 136, 143
sugar, 6, 148
supercomputer, 41
surface, 8, 19, 116, 139–140
    attachment, 116, 139–140
switch, 4, 150
synthesis, 149
    DNA, 11, 122
    protein, 10, 149

Taq, *see* enzyme, *Taq*
target sequence, 14–16, 18, 134
template, 10, 14–16, 124, 126–131, 133, 137, 144
terminator, 10, 148
thermal cycler, 15
Thermus aquaticus, 15
thymine, 6, 10, 131
tic-tac-toe, 145
tile, 61–63, 122, 124, 128–129, 144–145
time, 3, 36, 40–42, 45–46, 61, 64, 67–68, 73, 75–76, 91, 110
  constant, 50, 75–76
  exponential, 42, 72
  linear, 52, 76
  memory access, 1
  parallel, 50–55, 74–75, 91, 93–94, 100
  polylogarithmic, 71–72
  polynomial, 43, 55, 71–72, 115
  quadratic, 42–43, 77
  running, *see* running time
  unit, 52
transcription, 8, 10, 17, 109, 137, 148–149
transformation, 18, 95–96
transistor, 3
transition, 26–28, 60, 67–68, 109
transitive closure, 88–89

translation, 8, 10, 17, 109
  of algorithm, 71, 88, 91–92, 94–95, 97, 100, 102–104
tree, 33, 39, 66, 76
triple-crossover, 144
truth table, 25, 42
Turing machine, 27–29, 31, 44, 60–61, 63, 73, 77, 82, 109

universality, 29, 50, 63, 68, 91
uracil, 10

VLSI, 82
variable, 24–25, 32–33, 40–42, 48–49, 55, 77, 82–83, 93, 97, 136–141, 144
vector, 17–19, 125
vehicle, 18
virus, 17–18
volume, 3, 20, 45, 71–72, 88, 91, 94, 100, 107, 115, 118, 127
von Neumann, J., 1, 3, 148

whiplash PCR, *see* polymerase chain reaction (PCR), whiplash
"word processing", 109

XOR, 144–145

yeast, 7

Natural Computing Series

W.M. Spears: **Evolutionary Algorithms. The Role of Mutation and Recombination.**
XIV, 222 pages, 55 figs., 23 tables. 2000

H.-G. Beyer: **The Theory of Evolution Strategies.** XIX, 380 pages, 52 figs., 9 tables. 2001

L. Kallel, B. Naudts, A. Rogers (Eds.): **Theoretical Aspects of Evolutionary Computing.**
X, 497 pages. 2001

G. Păun: **Membrane Computing. An Introduction.** XI, 429 pages, 37 figs., 5 tables. 2002

A.A. Freitas: **Data Mining and Knowledge Discovery with Evolutionary Algorithms.**
XIV, 264 pages, 74 figs., 10 tables. 2002

H.-P. Schwefel, I. Wegener, K. Weinert (Eds.): **Advances in Computational Intelligence.
Theory and Practice.** VIII, 325 pages. 2003

A. Ghosh, S. Tsutsui (Eds.): **Advances in Evolutionary Computing. Theory and
Applications.** XVI, 1006 pages. 2003

L.F. Landweber, E. Winfree (Eds.): **Evolution as Computation.** DIMACS Workshop,
Princeton, January 1999. XV, 332 pages. 2002

M. Hirvensalo: **Quantum Computing.** 2nd ed., XI, 214 pages. 2004 (first edition
published in the series)

A.E. Eiben, J.E. Smith: **Introduction to Evolutionary Computing.** XV, 299 pages. 2003

A. Ehrenfeucht, T. Harju, I. Petre, D.M. Prescott, G. Rozenberg: **Computation in Living
Cells. Gene Assembly in Ciliates.** XIV, 202 pages. 2004

L. Sekanina: **Evolvable Components. From Theory to Hardware Implementations.**
XVI, 194 pages. 2004

G. Ciobanu, G. Rozenberg (Eds.): **Modelling in Molecular Biology.** X, 310 pages. 2004

R.W. Morrison: **Designing Evolutionary Algorithms for Dynamic Environments.**
XII, 148 pages, 78 figs. 2004

R. Paton[†], H. Bolouri, M. Holcombe, J.H. Parish, R. Tateson (Eds.): **Computation in Cells and
Tissues. Perspectives and Tools of Thought.** XIV, 358 pages, 134 figs. 2004

M. Amos: **Theoretical and Experimental DNA Computation.** XIV, 170 pages, 78 figs. 2005